Applications of Novel Scintillators for Research and Industry (ANSRI 2016)

AA006320

Journal of Physics: Conference Series Volume 763

Dublin, Ireland
11 – 13 May 2016

Editort:

Oliver J. Roberts
David Jenkins
Sheila McBreen
Brent Wagner

ISBN: 978-1-5108-3181-0
ISSN: 1742-6588

Printed from e-media with permission by:

Curran Associates, Inc.
57 Morehouse Lane
Red Hook, NY 12571

proceedings.com

Some format issues inherent in the e-media version may also appear in this print version.

Printed by Curran Associates, Inc. (2016)

For permission requests, please contact the Institute of Physics
at the address below.

Institute of Physics
Dirac House, Temple Back
Bristol BS1 6BE UK

Phone: 44 1 17 929 7481
Fax: 44 1 17 920 0979

techtracking@iop.org

Additional copies of this publication are available from:

Curran Associates, Inc.
57 Morehouse Lane
Red Hook, NY 12571 USA
Phone: 845-758-0400
Fax: 845-758-2633
Email: curran@proceedings.com
Web: www.proceedings.com

TABLE OF CONTENTS

In this volume of the Journal of Physics: Conference Series, we present the published proceedings from the "2nd Workshop on the Applications of Novel Scintillators for Research and Industry (ANSRI)." This workshop was held from the 11-13th of May, 2016, at University College Dublin, Ireland. The 12 proceedings from this meeting were all presented in oral or poster sessions. More details about this meeting, and the one before it, can be found on the workshop website at http://spacescience.ie/ansri2016/.

The scientific program of the 2nd ANSRI Workshop consisted of 7 invited seminars by world-renowned experts, 23 oral presentations and 6 poster presentations. The workshop was well attended by graduate and post-graduate students, post-docs and professors from academia, and junior staff scientists and researchers from industry. Participants from 14 countries attended the workshop, which covered a wide range of topics that characterise modern advances in the field of scintillator technology. The aim of this meeting was to establish a transfer of knowledge between researchers and industry experts from a wide range of application areas, in the dynamic field of scintillator technology. Moreover, the workshop also sought to help identify new pathways for future research, expanding upon existing networks and building upon current developments. This set of proceedings covers areas including the growth, production and characterisation of scintillators, along with their applications in various fields, which include: Gamma-ray Astronomy, Security and Defence, Beam monitoring, Nuclear Spectroscopy and Nuclear Astrophysics.

On behalf of the Local and the Scientific Organising Committees of the 2nd ANSRI Workshop, we are extremely grateful to all of the authors for providing us with details of their current research in these proceedings, and to the reviewers for their support. We would like to thank Professor Mark McConnell for giving a stimulating public talk during the workshop, and to the Institute of Physics Ireland group for supporting the event. We would also like to thank ORTEC/Ametek for sponsoring the workshop presentation prizes, and to the Science Foundation Ireland, Fáilte Ireland, SCIONIX, and the Institute of Physics for supporting and sponsoring the workshop. We also thank University College Dublin for use of the venue and support during the workshop. Finally, we would like to thank members of the Local and Scientific Organising Committees for their help in making this workshop meeting and the series, a success.

Dr. Antonio Martin-Carrillo & Dr. Isaac Tobin

On behalf of the Local Organising Committee

Dr. Oliver J. Roberts

Workshop Coordinator

Editors of the Proceedings:

Oliver Roberts (UCD, Ireland)
David Jenkins (Uni. of York, UK)
Sheila McBreen (UCD, Ireland)
Brent Wagner (Georgia Tech., USA)

Local Organising Committee:

Antonio Martin-Carrillo - **Co-Chair**
Isaac Tobin - **Co-Chair**
Oliver Roberts - **Coordinator**
Lorraine Hanlon
Sheila McBreen
David Murphy
Alexey Uliyanov

Scientific Organising Committee:

Oliver Roberts (UCD, Ireland) - **Chair**
Jarek Glodo (RMD, USA)
Lorraine Hanlon (UCD, Ireland)
David Jenkins (Uni. of York, UK)
Sheila McBreen (UCD, Ireland)
Marek Moszynski (NCBJ, Poland)
Francesco Quarati (T.U. Delft, The Netherlands)
Anant Setlur (General Electric, USA)
Kenneth Stanton (UCD, Ireland)
Brent Wagner (Georgia Tech., USA)

Journal of Physics: Conference Series **763** (2016) 011002 doi:10.1088/1742-6596/763/1/011002

Peer review statement

All papers published in this volume of *Journal of Physics: Conference Series* have been peer reviewed through processes administered by the proceedings Editors. Reviews were conducted by expert referees to the professional and scientific standards expected of a proceedings journal published by IOP Publishing.

Optical characterization of ion-doped crystalline and glassy matrices operating under hostile environmental conditions

S Baccaro and A Cemmi

ENEA Fusion and Technology for Nuclear Safety and Security Department, Casaccia R.C., Via Anguillarese 301, S.Maria di Galeria, 00123 Rome, Italy

stefania.baccaro@enea.it

Abstract. The aim of this work is to give an overall picture of the activity on gamma radiation effects in the field of scintillators (crystals, glasses) with specific mention to the role of ions doped in different crystalline and glassy matrices. Interesting results were obtained in terms of radiation hardness improvement and of physical-chemical properties modification as a function of the nature of dopant (i.e. rare earth as well as metallic ions) and of the scintillating host glassy and glass-ceramic matrix (silicate, phosphate, borate, oxyfluoride and mixed oxides glasses). The research activities were carried out at the ^{60}Co gamma Calliope plant, a pool-type irradiation facility located at the Research Centre ENEA-Casaccia (Rome). Since the eighties, the Calliope facility is deeply involved in radiation processing research on materials and on devices to be used in hostile radiation environment such as nuclear plants, aerospace and High Energy Physics experiments, in the framework of international projects and collaboration with industries and research institutions.

1. Scintillating crystals and glasses

Several experimental studies worldwide, are currently in progress to try and determine the effect of doping ions in both crystalline and glassy matrices. The choice of dopants needs to be selected carefully, as each dopant retains its intrinsic optical, luminescent and radiation resistant properties, when incorporated into the host matrices. This makes the doped-material suitable for a wide range of interesting and different applications.

A good scintillator is a material able to convert as large a fraction as possible of the incident radiation energy in the visible range to prompt fluorescence (occurring in times of the order of about 10^{-7} s), minimising the generally undesirable contribution for phosphorescence (occurring in times of the order $>>10^{-7}$ s) [1][2][3].

Scintillation detector consists of a scintillator (phosphor) material followed by an optional relay element and a photodetector. Wide band-gap materials are employed to convert X-rays to UV/visible photons. It is challenging to find dense, fast and radiation hard scintillator useful in high energy physics, astrophysics, medicine and other applications [4][5][6]. Currently, various scintillators such as NaI:Tl, CsI:Tl, BGO, BaF_2, $PbWO_4$, YAG, LuAG are used for many applications in the field of research, medicine and industry [3] [7] [8] [9].

The entire scintillation conversion can be divided into three processes: conversion, transport and luminescence. Conversion process involves an interaction of high-energy photon with the material lattice: if the radiation energy is around 1 MeV, the photoelectric and Compton scattering effects mainly occur [10]. As a result, many electron-hole pairs are created and thermalized in the conduction and valence bands. During the transport stage, these charge carriers migrate through the lattice. Their capture at trapping levels within the material forbidden gap delays migration and energy losses due to non-radiative recombination, which may occur. Finally, the luminescence stage involves radiative recombination of the electrons and holes trapped at the luminescence centre. The overall scintillation process efficiency η (number of light photons emitted/MeV), is the product of three factors, β, S and Q, related to the above described processes, respectively. The first (β) indicates the fraction of electrons-holes pairs actually produced, during the energy absorption, relative to the total that could be formed if there were no other losses. The factor S indicates the fraction of the total electron-hole pairs that is transferred to the luminescence centres. The quantum efficiency Q represents the fraction of excited centres that actually re-emit scintillation photons (figure 1) [11] [12].

Variations in the lattice (like defects and impurities) create local electronic energy levels in the energy gap of the scintillating matrix. If these levels are unoccupied, holes and electrons moving in the conduction band may be trapped by metastable centres, like luminescence and quenching centres and traps (figure 1). In the luminescence centres the transition to the ground state is accompanied by photon emission, while in the quenching centres radiationless thermal dissipation of excitation energy may occur. Finally, traps have metastable levels from which the electrons may subsequently return to the conduction band by acquiring thermal energy from the lattice vibrations, or fall to the valence band by a radiationless transition.

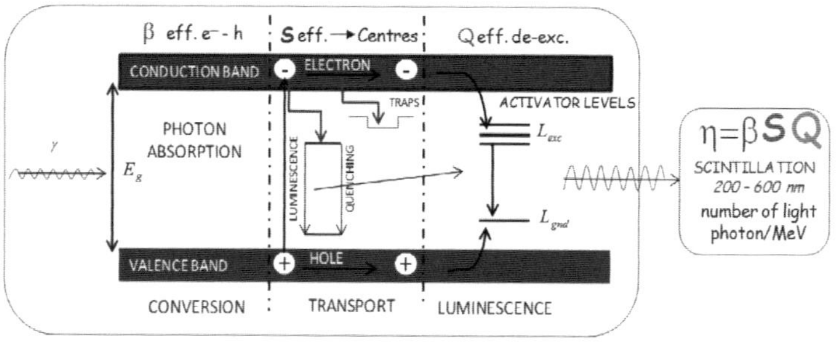

Figure 1. Schematic representation of scintillation process.

In order to enhance the probability of visible photon emission during the de-excitation process, small amounts of impurities called "activators" (usually rare-earth ions) are commonly added to scintillators, dispersed into the melting during the crystal growth. Activators, with an energy gap lower than that of the hosting crystal, create special sites in the lattice, not only modifying the normal energy band structure of the pure crystal, but also improving the performance of the scintillating matrix due to the interaction with the radiation induced charged defects (e, h). The transition between the activator excited and ground states (around 10^{-7}s) can occurs and the emission of a visible photon contributes to the scintillation process (figure 1). Increasing the ion dopant concentration results in a decrease of radioinduced absorption and of the light yield. For this reason, to improve the efficiency of a scintillator, a suitable dopant content must be chosen, being a compromise between the value of light yield and the resistance of the material to radiation.

Among the characteristics of scintillating materials, their radiation resistance is particularly important because high energy physics applications create rather severe environments, with high level of radiation as a result of particle collision. The radiation damage in solid inorganic matrices (ionic crystals, glasses, and various insulators and dielectric materials) generally results in a disruption of their crystalline structure. Any deviation from a perfect crystal lattice such as vacancies, interstitials, impurity atoms, thermal spikes and ionization effects (originated by the passage of charged particles or gamma rays through the material) can be considered as defects [13].

In case of gamma radiation, the initial stage of radiation induced damage is essentially a three-step process consisting of: i) creation of hot electrons and holes by the interaction with high energy photons; ii) their separation during subsequent cooling down and diffusion processes; iii) separate localization of both types of free carrier at suitable lattice sites (traps). Such sites might arise as a result of lattice distortions (dislocations, domain interfaces, etc.) or point defects (vacancies, impurity ions) or combinations of both of them. Colour centres might be created during phase iii). The lack of a negative ion in the lattice determines a local excess of positive charge which can act as a trap for electrons (F-centre). Electrons in colour centres have energetic levels different from those in the crystal and they can re-absorb the light emitted by scintillator in the visible region, reducing the scintillator performance. Radiation induced charge compensation could occur by oxidation and/or reduction reactions.

Scintillating materials can be investigated by steady state and time dependent characterizations. The former techniques include luminescence and transmission measurements, while the latter decay kinetics and scintillator light yield [11].

To investigate the damage caused by irradiation, the reduction of scintillator light output and optical transmission of the material are usually measured. The former is defined as the loss of the amount of light collected, after irradiation, by the photomultiplier when the scintillating material is exposed to incident radiation. For each type of scintillator it is necessary to measure, at different absorbed doses, the total radiation damage to the light output. The study of the radiation induced defects that involves modification of the optical transmission are usually evaluated by the radiation induced absorption coefficient μ, proportional to the ratio of the transmission before and after irradiation and normalized to the sample thickness.

2. Research and qualification activities at Calliope facility

Nowadays, ionizing radiation is being employed in a wide number of applications of great importance within both the scientific and industrial communities.

The Calliope gamma irradiation facility, located at the Research Centre ENEA-Casaccia (Rome), is deeply involved in qualification and research activities, in the framework of international projects and collaborations with industries and research institutions [14].

Many research activities are focused on the investigation of gamma irradiation effects on chemical and physical properties of different materials, such as radiation detectors, scintillating crystals and glasses, polymers, for several applications (nuclear plants, aerospace, High Energy Physics experiments) [3] [8] [9] [15] [16] [17] [18] [19] [20]. Material characterization and biological research has been carried out on cultural heritage archives and artifacts [21] [22].

Qualifications tests are mainly performed on electronic components and devices for application in hostile environments such as nuclear plant and aerospace and on concrete matrices for nuclear waste disposal and storage. As it will be shown, very special care is devoted to the qualification of electronic devices, which have to be tested according to MIL-STD-883 and/or ESA/SCC BASIC Specifications No.22900 procedures [23] [24] [25] [26] [27].

The Calliope facility is a pool-type irradiation facility equipped with a ^{60}Co γ source in a high volume (7x6x3.9 m^3) shielded cell. The present source has cylindrical geometry with the 48 source rods arranged in two concentric cylinders of about 20 cm outer radius and 26 cm height. The emitted radiation consists of two γ photons of 1.173 and 1.332 MeV emitted in coincidence, with a mean photon energy of 1.25 MeV. The maximum licensed activity for this plant is 3.7×10^{15} Bq (100 kCi).

Four kinds of dosimeters are used to determine the dose rate at which to irradiate the sample: Fricke solution, Red Perspex, alanine-ESR and TLD dosimeters [28].

An overview of some activities carried on at the Calliope facility related to doping and gamma radiation effects on crystal and glassy scintillators will be given in the following.

3. Scintillating crystals

3.1 Lead tungstate

Recently, $PbWO_4$ (PWO) has been chosen as a scintillating medium [7]. Position and shape of the peaks in scintillation spectra, decay kinetics, light yield and radiation resistance are the important parameters for any scintillation material. In as-grown PWO crystals, the emission spectrum is rather complicated and both regular lattice (420 nm "blue" emission) and defect-related (500 nm "green" emission) centres can contribute to the emission light [29] [30] [31] [32]. Under irradiation, a band at around 400 nm, responsible for the re-absorption of the scintillating light, is created. Together, two electron centres related to the radiation induced F^+ and F-centres, absorb scintillating light at 540 and 720 nm (figure 2). Considerable improvement of radiation resistance was found by doping PWO crystals with large and stable trivalent ions, such as La^{3+}, Lu^{3+}, Gd^{3+} and Y^{3+} [33] [34] [35] [36], as evident by transmittance and radiation induced absorption coefficient curves reported in figure 2 for La^{3+}. These positive effect can be explained considering that the Coulomb charge excess of trivalent ions at divalent lattice sites considerably reduces the possibility of temporary or stable localization of holes created during irradiation.

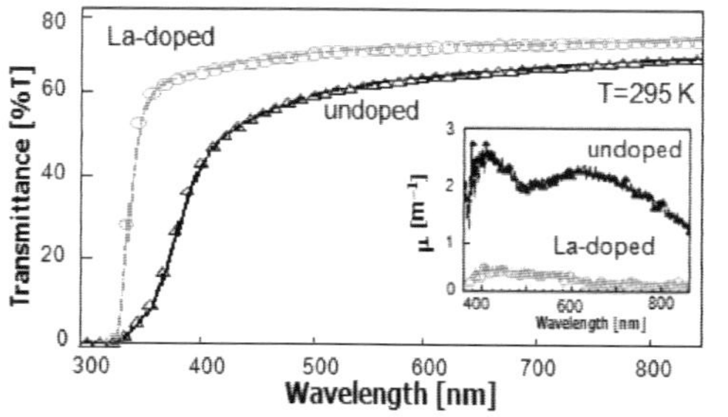

Figure 2. Transmittance curves of undoped and La^{3+}-doped PWO crystals. Inset: radiation induced absorption coefficient μ trends for the same samples [34].

Gadolinium doping appeared particularly interesting due to an additional effect, that consists in the improvement of the energy transfer efficiency in crystalline and glassy matrices. This sensitization mechanism was developed in some scintillating materials, with the addition of a high concentration (30%) of trivalent gadolinium ions. Gd^{3+} enables an efficient energy transfer to luminescence centres giving a sensible increase of the light emission, as confirmed by luminescence and decay kinetic measurements [3] [37] [38].

3.2 Cerium fluoride

Cerium fluoride (CeF_3) crystal was studied in the past to investigate its potential for fast scintillation and radiation hardness but is experiencing a rekindled interest as a scintillator for calorimetry at the LHC at CERN. For this reason, extensive researches were carried out to optimize the growth and post-preparation parameters and to obtain high reproducibility in term of optical and scintillation quality.

Pure CeF_3 crystals show emission peaks at 286 nm, 300 nm and 340 nm [7] [39]. Its optical and luminescence properties could be modified as a function of the amount of divalent ions (e.g. Ba^{2+}) added to the matrix, leading to a slight increase of the transmission band edge (figure 3(a)) and a shifting of the maximum of the emission peak to around 380 nm.

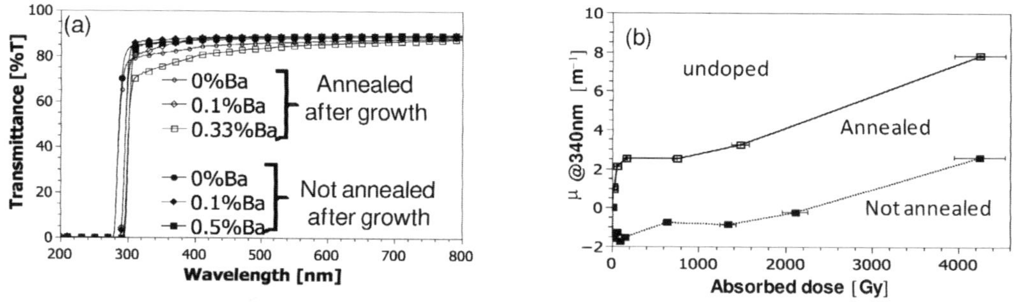

Figure 3. (a) Transmittance curves of undoped and Ba^{2+}-doped CeF_3 crystals annealed and not annealed after growth; (b) μ trends at the emission peak wavelength (340 nm) for the undoped samples annealed and not annealed after growth.

The results obtained after gamma irradiation on the same samples (data here not reported) show that the crystal doped with 0.5% Ba is characterized by the highest radiation hardness and stability.

Thermal treatment performed after the crystals growth limits also the radiation damage, as shown in figure 3(b) for undoped samples at the emission wavelength of 340 nm [40].

4. Glassy matrices

Over the last decade, glassy matrices have been of great interest in different technological applications. Due to their low cost, ease of production, chemical stability and light weight, these systems have been commonly used (in bulk or fibers form) as low and middle energy radiation detectors. Nevertheless, significant drawbacks are their low light yield, due to a low transfer efficiency of excitation energy from the matrix to luminescence centres and their low radiation hardness. Both effects are due to the presence of many trapping sites in the glassy matrices, where charge carriers can be trapped, giving rise to a non-radiative recombination. To improve the energy transfer mechanism, the sensitization effect due to the trivalent gadolinium ion, discovered for crystals (e.g. PWO), is successfully performed, as confirmed by photoluminescence and decay kinetics measurements [41].

4.1 Ce^{3+}-doped heavy germanate glasses

Glass matrices with density higher than 5 g/cm^3 containing GeO_2, BaO, Gd_2O_3 and doped with 0.5% of Ce_2O_3, were prepared by melting the glassy starting powder in air, inert atmosphere (N_2) and reducing atmosphere (N_2+graphite) [42]. The absorption edge of the glass melted in air is at longer wavelengths than that in reducing atmosphere: this behaviour can be explained because Ce^{3+} ions are easily oxidised to Ce^{4+} when exposed to air, causing the formation of broad absorption bands extending to the longer visible wavelength (figure 4) [43]. Ce^{3+} ions significantly interact with the hosting matrix and exert the red-shift effect on the ultraviolet and visible absorption position due to the 4f-5d electronic transitions, as shown in figure 5. Unlike exhibiting the monotonous deterioration effect on the glass matrix, radiation plays the radiation protection role on Ce^{3+}-doped glasses, depending on the glass melting atmosphere and on the radiation dose: this phenomenon is explained by the radiation induced reducing and oxidising mechanism due to the ability of Ce^{3+} and/or Ce^{4+} ions to capture both radiolytic holes and electrons [44].

The observed radiation induced effects seem to be relevant to an optimal radiation dose beyond which opposite influences occur. This is most likely associated with the double roles that radiation plays on glasses: when a limited radiation dose is applied, reducing and/or oxidising effects on glasses are dominant; as the radiation dose increases, concentration of radiation induced colour centres becomes higher and it finally exceeds what cerium ions can suppress. As a result, the radiation damage increases [45] [46].

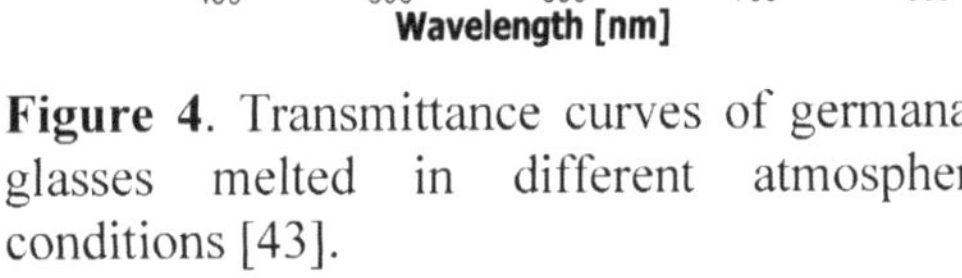

Figure 4. Transmittance curves of germanate glasses melted in different atmospheric conditions [43].

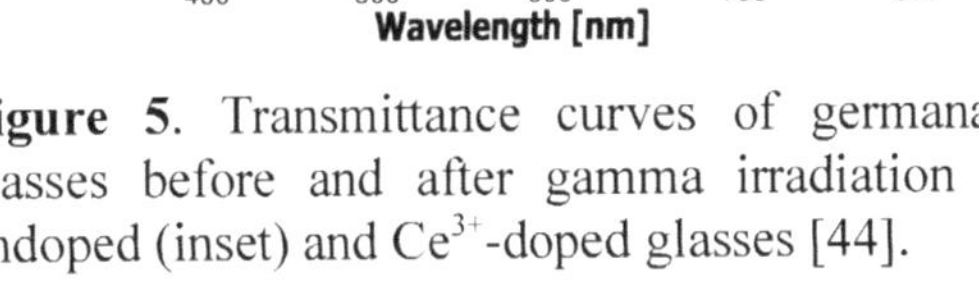

Figure 5. Transmittance curves of germanate glasses before and after gamma irradiation of undoped (inset) and Ce^{3+}-doped glasses [44].

4.2 Cerium and terbium doped silicate and phosphate glasses

Ce^{3+} ion is widely used as an activator in scintillating crystalline and glassy matrices. Depending on the hosting glass composition, its emission properties can vary. For example, cerium doping gives rise to a fast luminescence, with a typical decay time of about 20 - 50 ns, peaked at around 410 nm in silicate glasses and at around 350 nm in phosphate matrices [41] [47] [48] [49]. Considering silicate glasses, the increase of cerium doping (from 2% to 6%) induce a nearly negligible red-shift in the transmittance curves and very low modification after gamma irradiation up to 252 Gy in the range 360 - 470 nm (here not shown). The same matrix was also doped with 3% of Tb^{3+} ions: in this case, evident modification of the sample is clear in terms of the transmittance, when compared to a broad absorption band up to 500 or 600 nm. The radiation induced absorption coefficient μ for Ce^{3+} and Tb^{3+}-doped glasses monotonically decreases in the whole UV-VIS range and the cerium addition strongly improves the radiation hardness. This result is also confirmed comparing the μ value calculated at the emission wavelength correspondent to the single activator [49] [50] [51].

Similar investigation was carried out on Ce^{3+} and Tb^{3+} doped phosphate matrices containing 30 mol.% of Gd^{3+}. The comparison of the initial transmittance curves of phosphate glasses containing Ce^{3+}, shows a strong red-shift of the absorption band-edge with an increase of Ce^{3+} concentration, ranging from 325 nm for 3% of Ce^{3+} to 370 nm for 5% and 410 nm for 10% of Ce^{3+}. Such phenomenon limits the Ce^{3+} concentration that can be introduced in the glass matrix (3 - 4 mol.%), because it gives rise to a re-absorption of Ce^{3+} scintillating light (emitted at 350 nm), lowering the potential usage of these glasses as scintillating detectors. The transmittance curves before and after irradiation (10 Gy), in the case of cerium activated glasses show a slight transmittance decrease in the UV region of the spectrum, while an extended reduction from 250 to 550 nm of terbium is present. Finally, the comparison of μ coefficient for singly Gd^{3+}-doped and co-doped phosphate glasses with Ce^{3+} and Tb^{3+} shows that at the dopant proper emission wavelength the most radiation resistant sample is the terbium activated one (figure 6). In conclusion, these results indicate that doped-glasses radiation damage is strictly dependent on the matrices [48] [50] [51] [52].

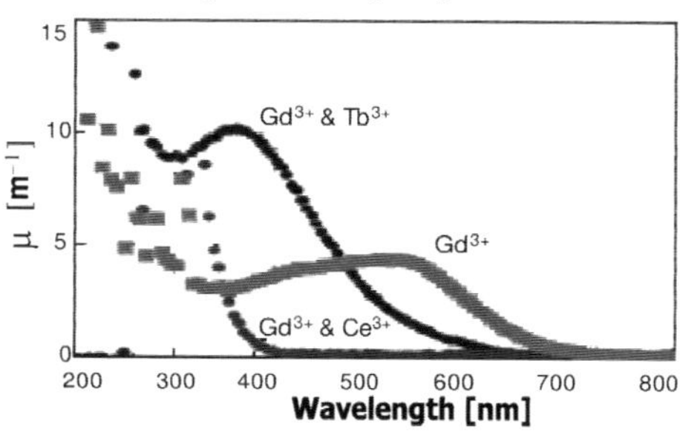

Figure 6. Radiation induced absorption coefficient plots for singly Gd^{3+}-doped and co-doped phosphate glasses with Ce^{3+} and Tb^{3+} [50].

4.3 Mixed oxide glasses containing ZnO as emission centre

Glassy matrices containing zinc oxide are studied for their luminescence properties and for applications as scintillating materials [53]. Interesting results were obtained by adding a high concentration (around 60 mol%) of ZnO in borosilicate glasses, without deteriorating the glass-forming ability, and investigating the thermal treatment effects on the photoluminescence emission spectra [54]. Despite the high concentration of zinc oxide, XRD spectra (here not shown) show that only ZnO crystalline aggregates are formed avoiding the ZnO bulk phase [54]. This evidence is also confirmed by the excitation and emission spectra reported in figure 7(a). When excited at the wavelength λ_{exc}=262 nm, the glass samples show a broad emission band at around 396 nm, while the emission band of ZnO bulk (range 500 - 600 nm) is absent. The emission band position is related to trapped excitons localized at the ZnO aggregate interface with the glass host. Compared with the untreated sample, the emission intensity of the heated glasses is sensibly modified (figure 7(b)), giving the most enhanced results for the glass treated at 350°C. This behaviour is reasonably explained with an energy transfer from the host glassy matrix to the luminescence ZnO aggregates.

The gamma irradiation effect was also investigated in a ZnO-based silicate matrix. In this system, SiO_2 acts as the network formers, BaO as network modifier leading to the formation of non-bridging oxygens, and Al_2O_3 (if present) as donors and glass network intermediate, having a tendency to substitute Si^{4+} with charge balance realized by Ba^{2+} atoms [55]. The ability of Al^{3+} ions dopant to increase the charged carrier concentration in the conduction band and broadening the band gap in the glassy matrix, is evident by the UV-shift of the cut-off wavelength in the transmission spectra. When aluminium oxide modifies the glass network, the non-bridging oxygens, which are known to be responsible for the colour centre formation, are bonded again, strengthening the glass structure and reducing the concentration radiation induced defects precursors in the matrix. The latter effect sensibly improves the radiation resistance of the glasses, as evident by the radiation induced absorption coefficient trends, after irradiation up to 262 Gy. While the matrix without aluminium shows a monotonic decrease of μ, the doped samples present a band peaked at around 380 nm.

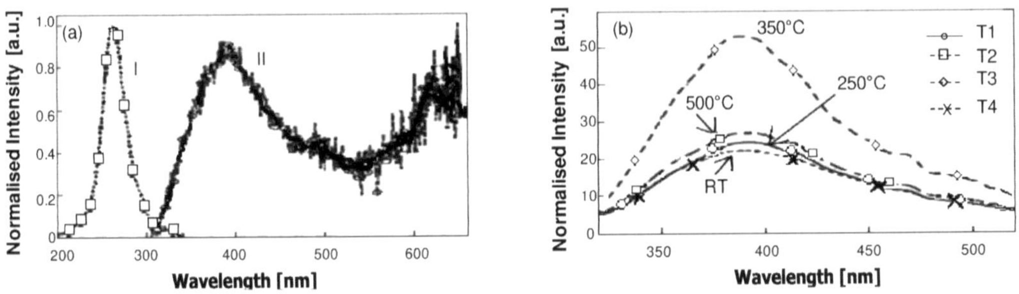

Figure 7. (a) Excitation (I) and emission (II) spectra of ZnO-borosilicate glasses heated at 350°C [54]; (b) Emission spectra of as prepared and heated ZnO-borosilicate glasses [54].

The same SiO_2-ZnO-BaO glassy matrix was also modified by the addition of barium fluoride, to investigate the effect of p-type F^- ions dopant on the luminescence and optical properties [53] [56] [57]. The BaF_2-free samples show an emission band centered at about 430 nm, as reported in figure 8. With the addition of F^-, the dominant emission band moves to shorter wavelengths at around 396 nm, approaching the excitonic peak emission of ZnO bulk (380 nm). A possible explanation is based on the possibility for F^- to replace the O^{2-} bonded to Zn^{2+} or some non-bridging oxygen, introduced by BaF_2, forming positive charged centres that allow the glass emission intensity increase.

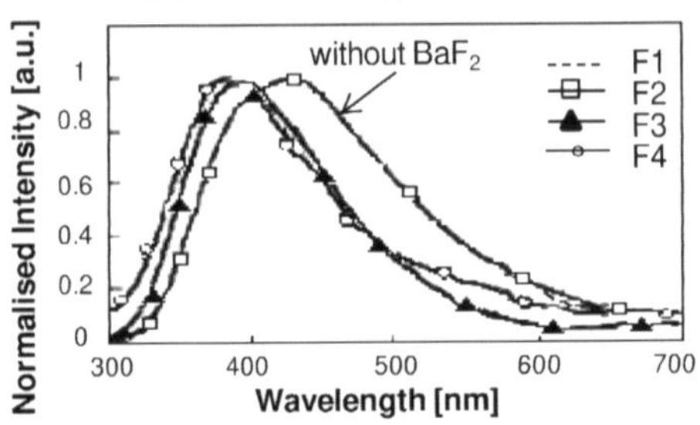

Figure 8. Emission spectra of SiO_2-ZnO-BaO (F1: $40SiO_2$-30ZnO-30BaO; F2: $40SiO_2$-35ZnO-25BaO) and SiO_2-ZnO-BaO-BaF_2 (F3: $40SiO_2$-30ZnO-25BaO-$5BaF_2$; F4: $40SiO_2$-30ZnO-20BaO-$10BaF_2$) glassy matrices [53].

5. Glass-Ceramics

Glass-ceramics are fine microstructured polycrystalline materials produced by heat treatment that induces a controlled crystallization (ceramization) of certain glass with specific composition. These materials are characterized by a high density (up to 99.5%) of small crystals embedded in a residual glassy phase. The starting glass (namely parent glass, formed by a standard glass-manufacturing process) and the final glass-ceramic do not present the same chemical composition and the ceramization process is usually rather critical to the attainment of acceptable and reproducible products. By designing the glass-ceramics composition, controlling the ceramization treatment and the resulting microstructure, an array of glass-ceramics materials with interesting and sometimes unusual combination of properties, are produced for several applications. Depending on their composition, glass-ceramics can combine a negative/zero/highly positive thermal expansion coefficient with optical transparency or opacity in the visible range, very high strength and toughness with translucency, chemical durability, biocompatibility and hardness for medical applications [58] [59]. Many glass-ceramics are also characterized by specific electric and magnetic features, low thermal conductivity, zero or very low porosity, thermal stability (400 - 1450°C) and mechanical properties [58] [60] [61].

The glass crystallization process, whose rate is temperature dependent, consists of reheating the parent glass above its transition temperature and occurs in two steps: a nucleation stage and a growth stage. During nucleation, that may take several hours, an extremely high density (10^{12}-$10^{15}cm^{-3}$) of small and stable crystalline phases nuclei are formed in preferred sites of the parent glass (interfaces within the parent glass or free surfaces). In most cases, nucleating agents (typically noble metals, fluorides, ZrO_2, Cr_2O_3, Fe_2O_3, TiO_2 or P_2O_5) are added to the parent glass composition to boost the nucleation process. The crystal-growth step involves the movement of atoms or molecules from the glass, across the glass-crystal interface, and into the crystals [61] [62] [63]. Several techniques allow the characterization of glass-ceramic materials. X-ray diffraction gives specific information regarding the crystallinity degree, the crystalline phases and sizes developed inside the glassy matrix and modified by heating. The distribution, composition and the dimension of the microcrystals are analyzed by the electron microscopy (SEM, TEM, EDX), while thermal analyses (DTA, DSC) allow us to determine the glass transition, crystallization and melting parameters as a function of the glassy system composition. Further characterization can be carried out by optical (UV-VIS, FTIR) and spectroscopic analyses (luminescence, ESR), physical measurements and mechanical tests.

An interesting investigation was carried out based on the results related to the BaF_2 addition effect on the silicate glasses photoluminescence (previously described). A SiO_2-ZnO-BaF_2 glassy matrix was studied as-prepared and after different heat treatments, with the aim to obtain the formation, inside the hosting matrix, of ZnO Quantum Dots (QDs). QDs consist of nanometer-sized particles and their electronic and optoelectronic properties can be tuned as a function of particle size and shape for a given composition. The emission peak of the parent glass, reported in figure 9(a), shows a maximum in the UV range at around 380 nm, when excited at 272 nm. If the sample is submitted to a different duration thermal treatment at 580°C, the emission band shifts to longer wavelength, due to the increase of ZnO QDs size (figure 9(b)).

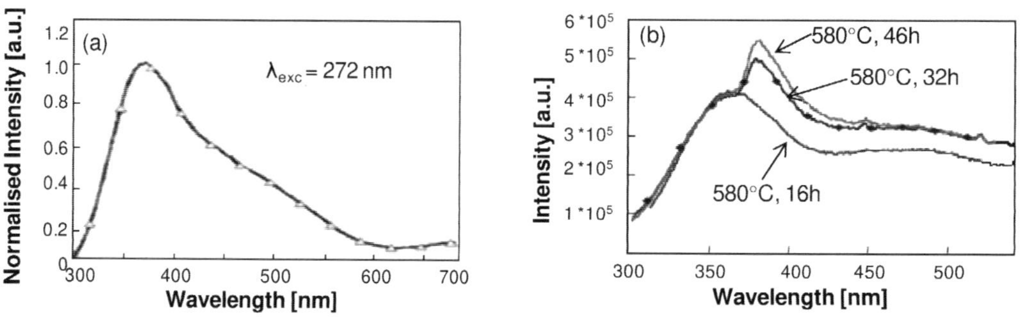

Figure 9. Emission spectrum of SiO_2-ZnO-BaF_2 parent glass: (a) as prepared; (b) after different thermal treatments [56].

6. Conclusion

As discussed in this work, research activities performed on crystal and glassy matrices for hostile environments, such as nuclear plants, aerospace and High Energy Physics experiments, highlighted the importance of the choice of suitable doping ions to obtain specific optical and luminescence properties on materials. Radiation resistance improvement is also strictly dependent on the crystal and glassy hosting matrices composition.

Acknowledgments

We are very grateful to Mr. G. Ferrara for his effective and continuous support during all the activities carried out at the Calliope facility and to INFN for the support in the framework of CMS, LUMEN, NEW LUMEN and BELLE II project.

References

[1] Rodnyi P A 1997 *Physical Processes in Inorganic Scintillators* (New York: CRC Press)
[2] Van Eijk C W E 2001 *Nucl. Instr. Meth. Phys. Res. A* **460** 1-14

[3] Nikl M, Bohacek P, Mihóková E, Baccaro S, Vedda A, Diemoz M, Longo E, Kobayashi M, Auffray W and Lecoq P 1999 *Nucl, Phys. B-Proc. Sup.* **78** 471-8

[4] Chatrchyan S *et al.* 2012 *Science* **338 (6114)** 1569-75

[5] Chatrchyan S *et al.* 2012 *Phys. Lett. B* **716 (1)** 30-61

[6] Chatrchyan S *et al.* 2012 *Phys. Lett. B* **710 (3)** 403-25

[7] Auffray E, Lecoq P, Korzhik M, Annenkov A, Jarolímek O, Nikl M, Baccaro S, Cecilia A, Diemoz M and Dafinei I 1998 *Nucl. Instr. Meth. Phys. Res. A* **402** 75-84

[8] Mihokova E, Nikl M, Pejchal J, Baccaro S, Cecilia A, Nejezchleb K and Vedda A 2007 *Phys. Status Solidi C* **4 (3)** 1012-1015

[9] Angelucci M *et al.* 2016 *Nucl. Instrum. Meth. A* **824** 678-80

[10] Price W J 1964 *Nuclear Radiation Detection* (New York: McGraw-Hill)

[11] Nikl M 2000 *Phys. Stat. Sol. (a)* **178** 595-620

[12] Blasse G 1994 Chem. Mat. **6** 1465-75

[13] Dienes G J and Vinegard G H 1957 *Radiation effects in solids Vol. II* (New York: Wiley)

[14] Baccaro S and Cemmi A 2011 *Proc.of SPIE - The International Society for Optical Engineering, August 21-25 San Diego (USA)*art. no. 81440F

[15] Baccaro S, Cemmi A, Di Sarcina I and Menchini F 2015 *Int. J. Appl. Glass Sci.* 1-7

[16] Baccaro S, Piegari A, Di Sarcina I and Cecilia A 2005 *IEEE Trans. on Nuclear Science* **52 (5)** 1779-1784

[17] Rossi P 2013 *Fusion Eng. Design* **88 (9-10)** 2084-2087

[18] Baccaro S 1996 *ACS Symposium Series* **620** 323-339 (Washington D.C.: ACS R. L. Clough and S. W. Shalaby)

[19] Aloisio A *et al.* 2015 *Frontier Detector for Frontier Physics 13th Pisa Meeting on Advanced Detectors May 24-30 La Biodola, Isola d'Elba (Italy)*

[20] Aloisio A *et al.* 2016 *Nucl. Instrum. Meth. A* Proc.-D-16_00219

[21] Baccaro S, Carewska M, Casieri C, Cemmi A and Lepore A 2013 *Polym. Degrad. Stabil.* **98 (10)** 2005-2010

[22] Adamo M, Baccaro S and Cemmi A, *ENEA Report RT/2015/5/ENEA*

[23] Ameel J *et al.* 2015 *J. Inst.* **10 (1)** 1-8

[24] Abbate C *et al.* 2015 *J. Inst.* **9 (2)** art. no. C02017

[25] Baccaro S *et al.* 1999 *Nucl. Instr. Meth. Phys. Res. A* **426 (1)** 206-11

[26] *Total dose steady-state irradiation test method*, ESA/SCC Basic Specification No. 22900, European Space Agency

[27] *Ionizing radiation (total dose) test procedure*, MIL-STD-883E, method 1019.4

[28] Baccaro S, Cemmi A, Ferrara G and Fiore S 2015 *ENEA Report RT/2015/13/ENEA*

[29] Baccaro S *et al.* 1998 *Proc. Int.Workshop on Tungtate Crystals* Oct.12-14 Rome (Italy) 129-37

[30] Nikl M *et al.* 2001 *Radiat. Meas.* **33 (5)** 705-708

[31] Baccaro S *et al.* 1999 *Radiat. Eff. Defects S.* **150 (1-4)** 15-9

[32] Kobayashi M *et al.* 1999 *Nucl. Instr. Meth. Phys. Res. A* **434 (2)** 412-423

[33] Nikl M *et al.* 1997 *Appl. Phys. Lett.* **71 (26)** 3755-7

[34] Baccaro S *et al.* 1997 *Phys. Status Solidi A* **160 (2)** R5-R6

[35] Nikl M et al. 1997 *J. Appl. Phys.* **82 (11)** 5758-62

[36] Baccaro S *et al.* 1999 *IEEE Trans. Nucl. Sci.* **46 (3 PART 1)** 292-5

[37] Baccaro S, Bohacek P, Borgia B, Cecilia A, Croci S, Dafinei I, Diemoz M and Vedda A *Phys. Stat. Sol. A* **164 (2)** R9-R10

[38] Baccaro S *et al.* 2000 *Phys. Stat. Sol. A* **179 (2)** 445-54

[39] Auffray E *et al.* 1996 *Nucl. Instr. Meth. Phys. Res. A* **383 (2-3)** 367-90

[40] Dissertori G, Nessi-Tedaldi , Baccaro S, Cemmi A, Fiore S, Fukuda K, Inui Y and Matsuo K 2015 *SCINT2015 13th Int. Conf. on Inorg.Scint. and their Appl.* June 7-12 Berkeley (USA)

[41] Nikl M, Nitsch K, Mihokova E, Solovieva N, Mares JA, Fabeni P, Pazzi G P, Martini M, Vedda A and Baccaro S 2000 *Appl. Phys. Lett.* **77 (14)** 2159-61

[42] Wang S, Chen G, Du Y, Wang H, Baccaro S and Cecilia A 2002 *Am. Ceram. Soc. Bull.* **81 (8)**

[43] Yang Y, Baccaro S, Cecilia A, Rao J, Zhang J, Xia F and Chen G 2005 *Nucl. Instr. Meth. Phys. Res. A* **537 (1-2 Spec. Iss.)** 435-8

[44] Chen G, Du Y, Baccaro S, Cecilia A, Mihokova E, Nikl M and Nitsch . 2001 *Am. Ceram. Soc. Bull.* **80 (4)** 107-10

[45] Baccaro S, Cecilia A, Chen G, Du Y, Nencini L and Wang S 2003 *Radiat. Eff. Defects S.* **158 (1-6)** 451-6

[46] Chen G, Baccaro S, Cecilia A, Du Y, Montecchi M, Nie J, Wang S and Zhang Y. 2003 *J. Non-Cryst. Solids.* **326-327** 343-7

[47] Nikl M *et al.* 2001 *Rad. Meas.* **33 (5)** 593-6

[48] Baccaro S *et al.* 2000 *J. Lumin.* **87** 673-5

[49] Baccaro S, Cecilia A, Montecchi M, Nikl M, Polato P, Zanella G and Zannoni R 2003 *J. Non-Cryst. Solids.* **315 (3)** 271-5

[50] Baccaro S, Cecilia A, Mihokova E, Nikl M, Nitsch K, Polato P, Zanella G and Zannoni R. 2002 *Nucl. Instr. Meth. Phys. Res. A* **476 (3)** 785-9

[51] Baccaro S, Cecilia A, Cemmi A, Chen G, Mihokova E and Nikl M 2001 *IEEE Trans. Nucl. Sci.* **48 (3)**

[52] Chen G, Baccaro S, Nikl M, Cecilia A, Du Y and Mihokova E 2004 *Am. Ceram. Soc. Bull.* **87 (7)**1378-80

[53] Qian G, Baccaro S, Falconieri M, Bei J, Cecilia A and., Chen G. 2008 *J. Non-Cryst. Solids* **354 (40-41)** 4626-9

[54] Chen G, Nikl M, Solovieva N, Beitlerova A, Rao J, Yang Y, Zhang Y, Jiang X and Zhu C 2004 *Rad. Meas.* **38** 771-4

[55] Cecilia A, Baccaro S, Chen G, Fang X, Donghui Z, Junbiao Z and Yunxia Y. 2005 *IEEE Trans. Nucl. Sci.* **52 (5 III)** 1792-4

[56] Nie J, Zhang J, Bei J and Chen G 2008 *J. Non-Cryst. Solids* **354** 1361-4

[57] Qian G, Nikl M, Bei J, Pejchal J, Baccaro S, Giorgi R, Cecilia A and Chen G 2007 *Opt. Mater.* **30 (1)** 91-4

[58] Ahrens B, Brand S, Büchner T, Darr P, Schoenfelder S, Paßlick C and Schweizer S 2011 *J. Non-Cryst. Solids* **357** 2264-67

[59] Yang Y, Zeng H, Ren J, Yuan S, Fan C, Sun L and Chen G 2012 *J. Am. Ceram. Soc.* **95 (7)** 2206-10

[60] Garai M and Karmakar B 2016 *J. Alloy Compd.* **678** 360-9

[61] Zanotto E D 2010 *Am. Ceram. Soc. Bull.* **89 (8)** 19-27

[62] Rawling R D, Wu J P and Boccaccini A R 2006 *J. Mater. Sci.* **41** 733-61

[63] Carter CB and Norton MG 2007 *Ceramic Materials-Science and Engineering* (Berlin: Springer) Chapter 26 pp.463-80

Research activity with different types of scintillation materials

K.-T. Brinkmann[1], A. Borisevich[2], S. Diehl[1], V. Dormenev[1], J. Houzvicka[3], M. Korjik[2], R.W. Novotny[1], H.-G. Zaunick[1] and S. Zimmermann[1]

[1]Justus-Liebig-Universität Giessen, II. Physikalisches Institut , Heinrich-Buff-Ring 16 D 35392 Giessen, Germany
[2]Institute for Nuclear Problems, Bobruiskaya11, 220030, Minsk, Belarus
[3]CRYTUR, spol. s.r.o., Palackeho 175, CZ-511 01, Turnov, Czech Republic

E-mail: stefan.diehl@exp2.physik.uni-giessen.de

Abstract. Nowadays there is a growing interest and demand in the development of new types of scintillation materials for experimental high energy physics. Future detector developments will focus on cheap, fast, and radiation hard materials, especially for application in collider experiments. The most recent results obtained by the Giessen group in close cooperation with colleagues from different institutes will be presented. The new start of the mass production of high quality lead tungstate crystals ($PbWO_4$, PWO) for electromagnetic calorimetry was started by the company CRYTUR (Turnov, Czech Republic). We will present a detailed progress report on the research program of lead tungstate performed in the last two years. The latest results in the development of LuAG:Ce, YAG:Ce and LYSO:Ce inorganic fibers, grown by the micro pulling down method and cut with the heated wire technique as well as new glass ceramics material $BaO*2SiO_2$ (DSB) doped by Ce and Gd will be presented. In addition, different samples of the organic plastic scintillator EJ-260 produced by the company Eljen Technology (Sweetwater, USA) have been characterized. The study has focused on the change of performance after irradiation with 150 MeV protons up to an integral fluence of $5 \cdot 10^{13}$ protons/cm^2 as well as with a strong ^{60}Co gamma-source accumulating an integral dose of 100 Gy.

1. A new production of $PbWO_4$ crystals

The lead tungstate ($PbWO_4$, PWO) scintillation crystal is one of the most widely used scintillation materials in electromagnetic calorimetry in modern accelerator experiments [1]. It combines a unique set of physical and scintillation properties, which allow the construction of an affordable compact electromagnetic calorimeters (EMC) of high quality. The first generation of mass production of PWO-I crystals was used in the electromagnetic calorimeter of the CMS detector at LHC. The technology of mass production was optimized to obtain detector units with high radiation hardness and less stringent requirements to the light yield [2] due to the high photon energies to be expected. The PANDA EMC will be based on an improved quality (PWO-II) for increased light yield and radiation hardness [1]. The operating temperature of the PANDA EMC will be -25° C that allows to increase the light yield by a factor of four by reducing the quenching of luminescence but blocks the statistical recovery of radiation damage during operation. Crystals for CMS and partly for the PANDA calorimeter were produced by the Czochralski method at Bogoroditsk Technological Chemical Plant (BTCP, Bogoroditsk, Russia) until its shut down some years ago. The company CRYTUR (Turnov, Czech Republic) has a good experience in the development and production of different types of inorganic oxide crystals and has initiated in 2014 a research program to develop the technology for mass production of PWO-II crystals in close collaboration with Giessen University (Germany) and RINP (Minsk, Belarus). The test results of one of the first PWO crystals with dimensions of 2x2x20 cm^3 are presented on Figs. 1 and 2. The longitudinal optical transmittance was measured using a Varian Cary 4000 spectrophotometer. Transversal optical transmittance was measured with a Hitachi 3200 spectrophotometer. The radiation hardness of the sample can be characterized by the radiation induced coefficient which is defined as $dk= Ln[T_b/T_a]/d$, where T_b and T_a are the optical transmittances before and after irradiation, correspondently, and d is the thickness of the PWO sample. The irradiation was performed at the Radiation Center of Justus-Liebig-University (Giessen, Germany) with a set of ^{60}Co γ–sources. The dose of 30 Gy corresponds to the value, which will be reached in the most forward part of the PANDA EMC after half year of the operating time. The light yield was measured using a calibrated Hamamatsu R2059-01 photomultiplier and a ^{137}Cs γ–source. The obtained results confirm already now that the technological approach of CRYTUR will allow to produce lead tungstate crystals with properties very close to the PWO-II specifications of the PANDA experiment at FAIR (Darmstadt, Germany). As a next step, a pre-production of crystals in PANDA geometry will be started and characterized using a set of additional and technically optimized furnaces.

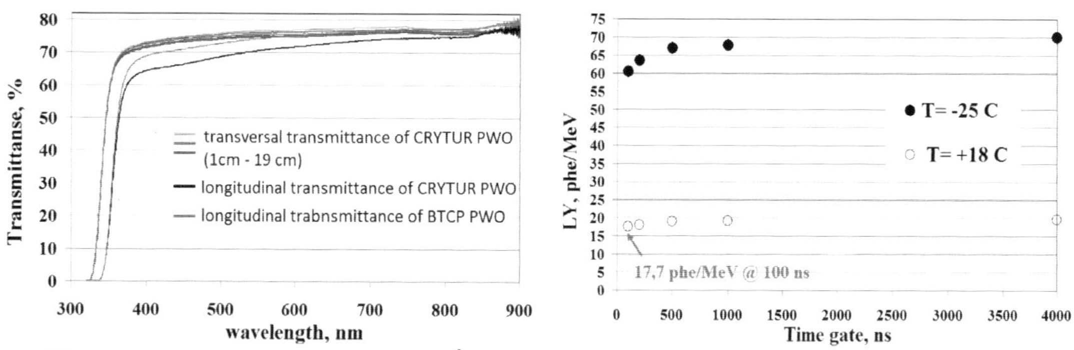

Figure 1. Test results of a 2x2x20 cm^3 PWO Crystal produced by CRYTUR: Longitudinal and transversal transmittance spectra in comparison with a BTCP type crystal (left figure). Light yield dependence on the integration time measured at different temperatures (right figure).

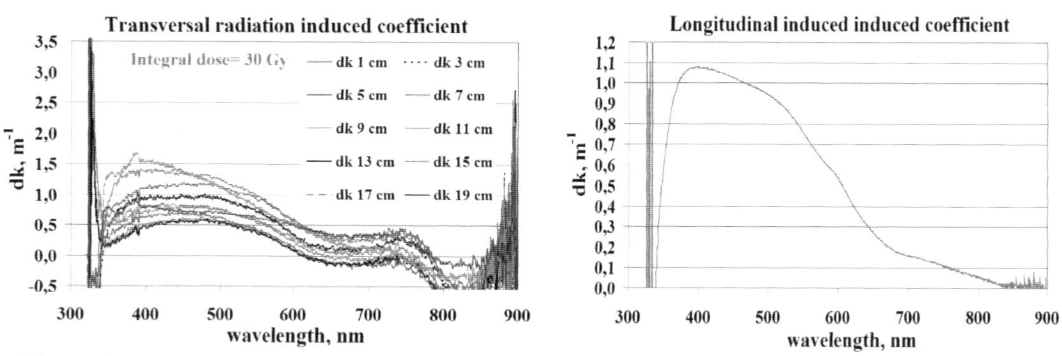

Figure 2. The transversal radiation induced coefficient measured at different positions of the 20 cm long crystal (left figure) compared to the longitudinal coefficient (right figure) after an integral γ-dose of 30 Gy.

2. Inorganic fibers made of LuAG:Ce, YAG:Ce and LYSO:Ce

Compared to conventional organic plastic scintillator fibers, inorganic fibers provide a series of advantages, like a significantly higher light yield, a more effective interaction with electromagnetic probes and a higher resistance to radiation damage. Due to these advantages and due to the high granularity provided by the fiber geometry, there are a lot of applications for these fibers in high

Journal of Physics: Conference Series **763** (2016) 012002 doi:10.1088/1742-6596/763/1/012002

energy physics experiments as well as in medical applications. Especially the implementation of new concepts in sampling calorimeters like dual readout calorimetry and the use in PET scanners are under investigation.

2.1. Fibers grown by the micro pulling down method

During the last years a R&D program for the optimization and characterization of fibers grown by the micro pulling down method has been performed. The crystal melt is heated up in a crucible with a radio frequency heater and the fiber is pulled down on a seed through a small opening in the bottom of the crucible. The quality of the fibers produced by Fibercryst (Lyon, France) could be increased significantly by an optimization of the pulling speed, the temperature gradient and the Cerium concentration and distribution within the fiber [3]. Fig. 3 shows the improvement of the attenuation coefficient of the scintillation light in typical fibers of 1-2 mm diameter, starting at approximately 1.0 cm^{-1} in 2009 down to 0.02 cm^{-1} in 2014 [3,4,5].

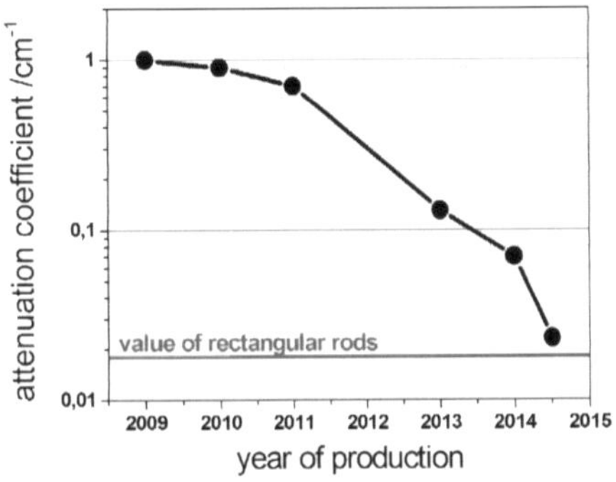

Figure 3. Improvement of the attenuation coefficient of LuAG:Ce fibers grown by the micro pulling down method over the last years. The green line shows the attenuation value measured for a rectangular rod cut from a large crystal and polished on all side faces [3].

2.2. Fibers cut with the heated wire method

As an alternative to fibers grown by the micro pulling down method, fibers cut by the heated wire technique from a conventional crystal, produced by the company CRYTUR (Turnov, Czech Republic) are currently under investigation. Fig. 4 shows several 1.0×1.0×100 mm^3 large rectangular LuAG:Ce fibers.

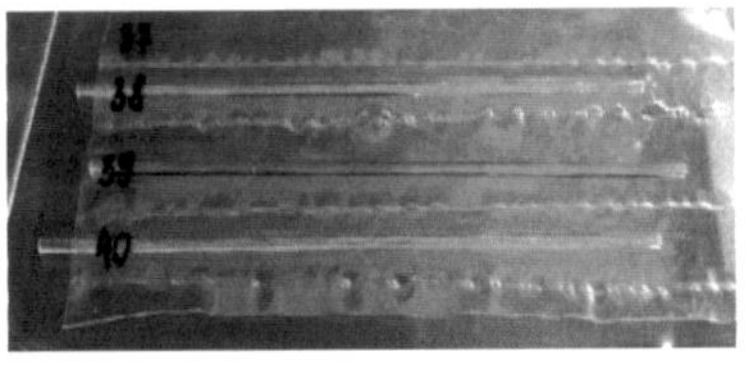

Figure 4. Picture of three 1.0 × 1.0 × 100 mm^3 large rectangular LuAG:Ce fibers cut by the heated wire technique.

The position dependent light yield of these fibers, measured by a SiPM with a cell size of 50 μm is shown in the left part of figure 5, while the right part of the figure shows a comparison to fibers grown by the micro pulling down method. The fibers show typical attenuation coefficients between 0.03cm^{-1} and 0.06 cm^{-1} which is comparable to typical Ø 2 mm fibers and even slightly better than the values of Ø 1 mm fibers grown by the micro pulling down method. The detected light yield, measured with α-particles of an ^{241}Am source, shows comparable values for both fiber types.

As an application of the rectangular fibers, a first sampling calorimeter prototype, consisting of four modules with dimensions of 13×13×100 mm^3 has been tested with a tagged photon beam in the energy range from 50 MeV up to 800 MeV in the A2 hall of the MAMI accelerator in Mainz. The single modules consisted of a tungsten array with 1.0×1.0×100 mm^3 large rectangular LuAG:Ce fibers inserted in the holes between the 1 mm thick tungsten layers. The readout of the modules, shown in Fig. 6 has been performed with PMTs with a bi-alkali photocathode coupled to the rear face of the modules via a light guide. The achieved relative energy resolution for the single modules, with a beam interacting centrally in each module is shown in the right part of Fig. 6.

Journal of Physics: Conference Series **763** (2016) 012002 doi:10.1088/1742-6596/763/1/012002

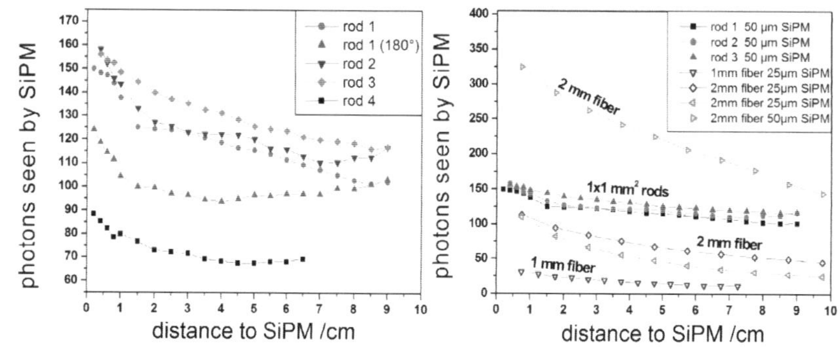

Figure 5. Light yield of different samples of 1×1×100 mm^3 large rectangular LuAG:Ce fibers cut by the heated wire technique (left). Right: comparison of the fibers (rod 1-3) with fibers grown by the micro pulling down method.

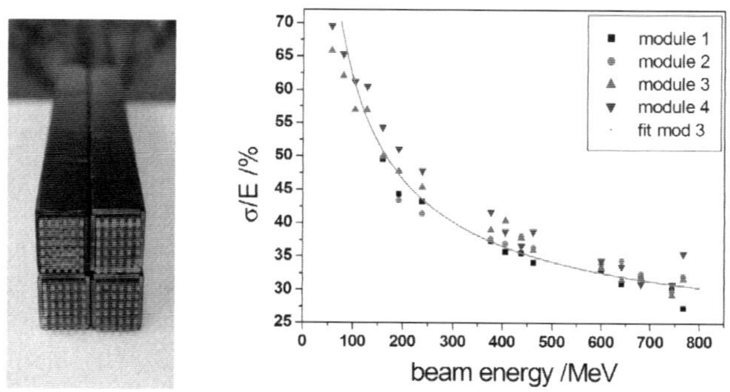

Figure 6. Picture of a sampling calorimeter module equipped with rectangular 1.0×1.0×100 mm^3 large rectangular LuAG:Ce fibers (left) and the energy resolution of the 4 modules measured with a tagged photon beam.

3. DSB:Ce - A new scintillator material

Recently, a new scintillation material, DSB:Ce^{3+}, was announced [6]. It can be produced as glass or nano-structured glass ceramics with application of standard glass production technology with successive thermal annealing. When doped with Ce $^{3+}$, the material can be applied as scintillator. The light yield of luminescence is near 100 ph/MeV. Un-doped material has a wide optical window from 4.5eV and can be applied to detect Cherenkov light. Several samples with dimensions of 15x15x7 mm^3 have been tested for damage effects under irradiation with gamma-quanta and 150 MeV protons. Results of the tests of one of these samples are presented in Fig.7. No significant damage after both type of irradiations was observed.

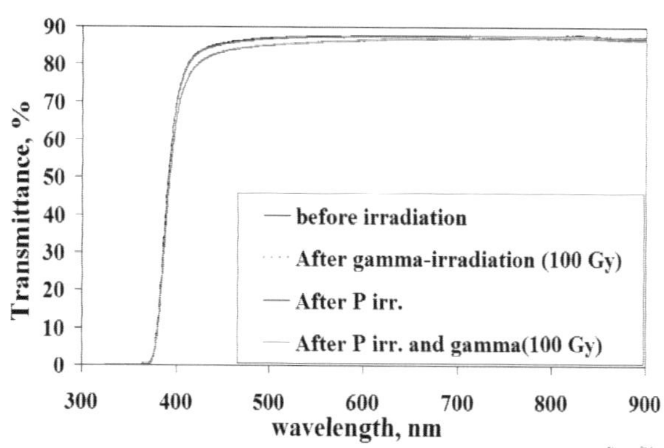

Figure 7. Optical transmittance spectra of the small sample of DSB:Ce^{3+} material before and after irradiations with 150 MeV protons and low energy γ-rays

The first full sized (12.5 cm length) sample of DSB:Ce^{3+} was produced at the end of 2015. A relatively large amount of uniformly distributed macro-defects were observed. The properties of the sample were measured in the same manner. Results of the light yield measurements at different temperatures as well as transversal and longitudinal optical transmittances are presented in Fig. 8. The relatively large differences of the slow component at different temperatures have to be further studied. The transversal optical transmittance shows satisfied uniformity along the sample. Relatively low

Journal of Physics: Conference Series **763** (2016) 012002 doi:10.1088/1742-6596/763/1/012002

value of the longitudinal transmittance in comparison with transversal values can be explained by Rayleigh scattering on the macro-defects. Therefore, further optimization of the production technology of large volume samples of DSB:Ce^{3+} is needed.

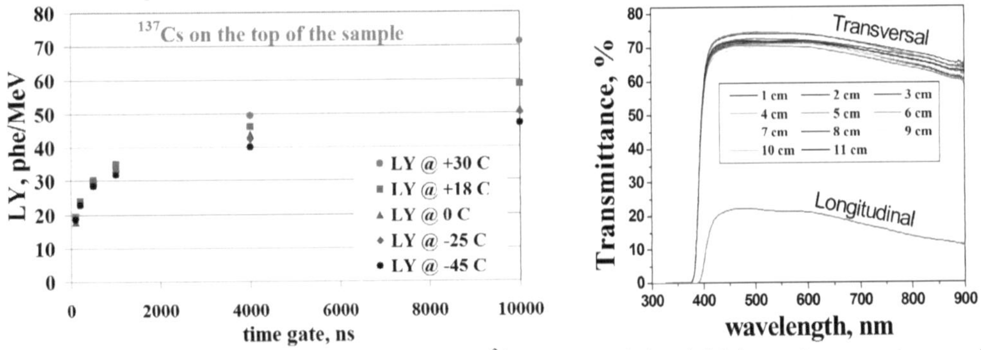

Figure 8. Test results of a 12.5 cm long DSB:Ce^{3+} sample. Light yield dependence on integration time at different temperatures (left). Longitudinal and transversal transmittance spectra (right).

4. Radiation hardness of the organic scintillator EJ-260

Plastic scintillators play a significant role in the construction of large area detectors in high energy physics experiments. Future concepts for detectors at HEP experiments, particularly at collider facilities, will require an unique combination of the material features and affordable price. Crucially important will be a minimal level of radiation damage effects under the electromagnetic part of the ionizing radiation as well as due to high energy hadrons. It requires a tolerable deterioration of the optical transmittance, a moderate level of the damage of the scintillation mechanism itself and a minor contribution of the radio-luminescence due to radio-nuclides which might be generated by nuclear reactions in the material of the detector. Here we report on first results of the study of commercially produced plastic scintillation material EJ260 (Eljen Technology, Sweetwater, USA). This bright, fast, green light emitting scintillator has properties similar to other products available on the market such as BC428 and NE103, respectively. The samples have approximate dimensions of 5.0×5.0×1.26 cm^3. The two large opposite sides of each sample were optically polished.

4.1. Irradiation with ^{60}Co gamma-quanta

The original surfaces of the sample 081015Q02 were unfortunately mechanically damaged during the preparations for the tests. Therefore, we tried to perform optical re-polishing procedure. The thickness of all samples after the polishing procedure was reduced to the value of 9.8 mm.

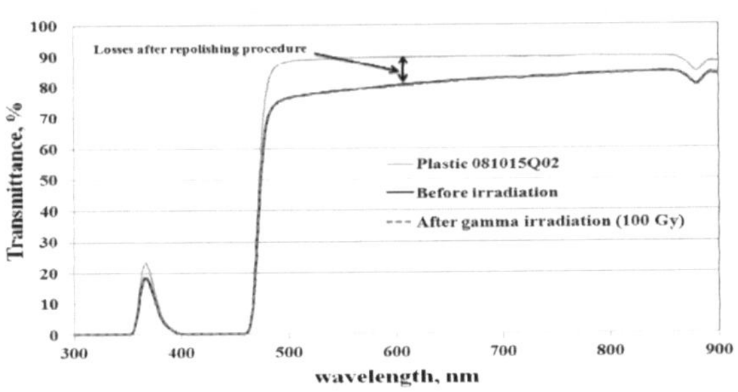

Figure 9. Optical transmittance spectra of a sample 081015Q02. The green curve is the distribution obtained before the mechanical damage of the surfaces, black and red curves are data measured after the re-polishing procedure and before and after irradiation with γ-rays, respectively.

Measurements of the optical transmittance of the sample after the repolishing procedure and before and after irradiation with the ^{60}Co γ-source (integral dose =100 Gy) were performed. The results in comparison to data obtained before the repolishing are presented in Fig. 9, which shows a significant loss of the optical transmittance after the repolishing procedure but no considerable losses due to γ-irradiation.

4.2. Irradiation with 150 MeV protons

For irradiation tests with a 150 MeV proton beam at KART-KVI (Groningen, Netherlands), one sample was cut in two parts with similar dimensions (2.5x5.0x1.26 cm^3). One piece of the scintillator was placed in front of a 4.5 mm thick lead block and was numbered "1". The second piece was placed behind and numbered as "2". The transversal size of the lead block was significantly larger than both scintillators. The assembly of two scintillator layers with a lead sheet in between was chosen to simulate a single section of a sampling calorimeter. The assembly "scintillator 1/lead/scintillator 2" was irradiated with 150 MeV protons impinging perpendicular to the front face of scintillator 1 with an integral fluence of $5 \cdot 10^{13}$ protons/cm^2. The scintillator samples were separated from the lead block six hours after the end of irradiation. Since the samples showed a relatively high level of radiation induced activity they had to be stored for two months at a temperature below -25°C. Receiving the samples back, the optical transmittance was measured again as well as after an additional irradiation with γ-rays of a ^{60}Co-source with an integral dose of 100 Gy (1.5 Gy/min.). The transmittance spectra before and after the irradiation procedures as well as spectra of the radiation induced coefficient of both samples after irradiation with 150 MeV protons and additional γ-rays are presented in Fig. 10.

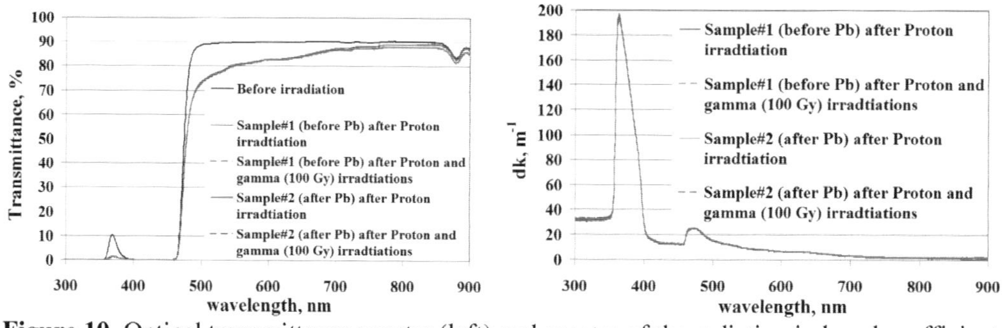

Figure 10. Optical transmittance spectra (left) and spectra of the radiation induced coefficient (right) of both samples before and after irradiations with 150 MeV protons and low energy γ-rays.

The radiation induced spectra do not indicate any difference with respect to the positioning of the scintillator relative to the lead block. Secondary and forward emitted reaction products with lead do not cause additional damage in the sample placed behind the absorber block. The irradiation with a dose of 100 Gy of γ-rays does not induce a visible additional damage effect in any of the samples.

References

[1] PANDA collaboration, ***Technical Design Report for PANDA Electromagnetic Calorimeter (EMC)*** (2008), http://arxiv.org/abs/0810.1216v1.

[2] E. Auffray et al., "Status of PWO crystal production for the electromagnetic calorimeter of CMS and of its construction," *Nucl. Instr. Meth. A*, vol. 537, pp. 373-378 (2005).

[3] S. Diehl, R. W. Novotny, N. Aubry, S. Faraj, G. Ferro and M. Kavatsyuk, ***Characterization and Optimization of New High-Quality Inorganic Fibers Made of LuAG:Ce and LYSO:Ce***, Conference Record 2014 IEEE Nucl. Sci. Symp. and Med. Imag. Conf. (2014).

[4] S. Diehl, R. W. Novotny, N. Aubry. S. Faraj, G. Ferro,***"Characterization and Applications of New High Quality LuAG:Ce and LYSO:Ce Fibers"***, Journal of Physics: Conference Series 587, 01206 (2015), doi: 0.1088/1742- 6596/587/1/012067

[5] S. Diehl, R. W. Novotny, N. Aubry. S. Faraj, G. Ferro, ***"Development and Characterization of Inorganic Scintillating Fibers Made of LuAG:Ce and LYSO:Ce"***, IEEE Transactions on Nuclear Science 61.1: 353-361 (2014).

[6] K. -T. Brinkman, A. Borisevich, V. Dormenev, et al., ***Radiation damage and recovery of medium heavy and light inorganic crystalline, glass and glass ceramic materials after irradiation with 150 MeV protons and 1.2 MeV gamma-rays***, Conference Record 2014 IEEE Nucl. Science Symp. and Med. Imag.Conf. (2014).

Brilliant gamma beams for industrial applications: new opportunities, new challenges

V Iancu[1], G Suliman[1], G V Turturica[1], M Iovea[2], I Daito[3], H Ohgaki[3], C Matei[1], C A Ur[1] and D L Balabanski[1]

[1] Extreme Light Infrastructure-Nuclear Physics/ Horia Hulubei National Institute for R&D in Physics and Nuclear Engineering, 30 Reactorului Street, Bucharest-Magurele, 077125 Romania
[2] Accent Pro 2000, s.r.l., Nerva Traian 1, K6, Apt 26, Bucharest, 031041 Romania
[3] The Institute for Advanced Energy Kyoto University, Uji, Kyoto 6110011, Japan

E-mail: violeta.iancu@eli-np.ro

Abstract. The Nuclear Physics oriented pillar of the pan-European Extreme Light Infrastructure (ELI-NP) will host an ultra-bright, energy tunable, and quasi-monochromatic gamma-ray beam system in the range of 0.2–19.5 MeV produced by laser-Compton backscattering technique. The applied research program envisioned at ELI-NP targets to use nuclear resonance fluorescence (NRF) and computed tomography to provide new opportunities for industry and society. High sensitivity NRF-based investigations can be successfully applied to safeguard applications and management of radioactive wastes as well as to uncharted fields like cultural heritage and medical imaging. Gamma-ray radioscopy and computed tomography performed at ELI-NP has the potential to achieve high resolution in industrial-sized objects provided the detection challenges introduced by the unique characteristics of the gamma beam are overcome. Here we discuss the foreseen industrial applications that will benefit from the high quality and unique characteristics of ELI-NP gamma beam and the challenges they present. We present the experimental setups proposed to be implemented for this goal, discuss their performance based on analytical calculations and numerical Monte-Carlo simulations, and comment about constrains imposed by the limitation of current scintillator detectors. Several gamma-beam monitoring devices based on scintillator detectors will also be discussed.

1. Introduction

The Extreme Light Infrastructure-Nuclear Physics (ELI-NP) facility under construction in Romania will host two state of the art systems, a high power laser system and a high brilliance gamma beam system [1]. The research program of ELI-NP focuses on laser and gamma driven nuclear physics as well as on applied research and development. A great variety of interdisciplinary scientific projects with societal and technological impact like industrial applications, medical applications and material applications are under development at ELI-NP[2, 3, 4, 5]. All the applications envisioned at ELI-NP exploit the unique features of the future laser and gamma beams to deliver new opportunities for society and industry.

The ELI-NP gamma beam is produced through Compton backscattering of laser light of an accelerated electron beam [6]. The laser light is delivered by two high power and high quality J-class lasers. The electron acceleration is based on a two-stage warm LINAC accelerator. The first stage of acceleration will deliver electrons of energies up to 300 MeV which will be deflected to a low energy interaction point generating gamma rays up to 3.5 MeV. The second stage will accelerate electrons up to 700 MeV, which will be deflected to a high energy interaction point to generate gamma rays up to 19.5 MeV. The main characteristics of the LCS gamma beam are presented in Table 1.

Table 1. Design characteristics of the ELI-NP LCS gamma beam [6].

Energy (MeV)	2.0	3.45	9.87	19.5
Nr of photons/pulse within FWHM	$1.2 \cdot 10^5$	$1.1 \cdot 10^5$	$2.6 \cdot 10^5$	$2.5 \cdot 10^5$
Nr of photons/s within FWHM	$4.0 \cdot 10^8$	$3.7 \cdot 10^8$	$8.3 \cdot 10^8$	$8.1 \cdot 10^8$
Source rms size (μm)	12	11	11	10
Source rms divergence (μrad)	140	100	50	40
Macro repetition rate (Hz)	100	100	100	100
Nr of pulses per macropulse	32	32	32	32
Pulse to pulse separation (ns)	16	16	16	16
Pulse length rms (ps)	0.92	0.91	0.95	0.90

As a result of the electron beam structure and the laser recirculation time the photons are emitted in very high intensity and in short pulses [6]. The macropulse repetition rate is 100 Hz. Each macropulse consists of 32 pulses separated by 16 ns; each pulse is 1 ps long. The number of photons within a pulse is around 10^5, which build up to 10^8 photons per seconds. The high intensity of the gamma beam constitutes a great opportunity for research and industry but also challenges for the current detection technology.

The gamma beam industrial applications that we envision are mainly nondestructive inspections of objects, which are grouped in active interrogations based on nuclear resonance fluorescence (NRF) and gamma ray radioscopy and computed tomography (CT). Here we describe the experimental setups under construction at ELI-NP, discuss their future performance based on analytical calculations and numerical simulations and comment on the challenges they present. In addition we present several gamma beam monitoring instruments based on scintillators that will determine the operational conditions and characterize the gamma beam.

2. NRF-based applications

Nuclear resonance fluorescence, a nuclear process that yields characteristic signatures for every nuclei, became an attractive method to use for non-destructive material composition studies and has already been tested in fields related to safeguards and waste management [7, 8, 9]. Prospective uses of NRF for uncharted fields like cultural heritage and medical diagnostics have been proposed and are yet to be demonstrated [10].

The setups dedicated to NRF-based applications to be developed at ELI-NP employ both the scattering and the transmission methods [2] and rely on the availability of a high-efficiency detector array. The ELI-NP array of detectors (ELIADE) will consist of eight segmented clover detectors and four $LaBr_3$ detectors placed on two rings at 90° and 135°. ELIADE is designed to measure gamma rays for the entire gamma-beam energy range and is optimized for high signal to background ratio [11]. The NRF transmission setup is sketched in Fig. 1. Here, ELIADE is placed downstream of the investigated object and measures the scattering of the transmitted beam of a witness foil, which is an enriched sample of the material of interest. A decrease in the NRF signatures detected at the resonance energy is correlated with the presence of a particular isotope in the object. A detector placed in the beam, downstream of the witness foil, measures the transmitted radiation.

Estimates of the NRF count rates for both the scattering and the transmission methods at ELI-NP suggest an improvement of around three orders of magnitude at ELI-NP compared to the available gamma-ray facilities [2]. A simulation study based on the NRF transmission method was carried out to obtain the transmission factor of the γ-rays of 2176 keV (excitation energy of ^{238}U) through an object that contains both ^{235}U and ^{238}U [12]. A clear distinction between the ^{235}U and ^{238}U distributions in the investigated object was made when the transmission factor of on-resonance γ-rays normalized by that of off-resonance γ-rays is used for the CT image reconstruction. Thus, NRF used in combination with computed tomography at ELI-NP can produce isotope-specific maps at single resonance level.

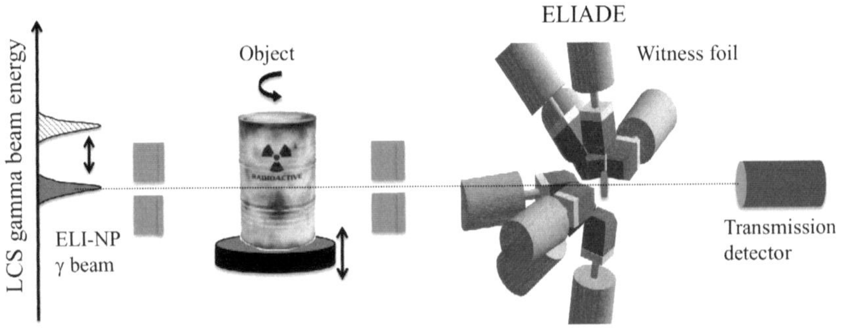

Figure 1. Schematic drawing of the transmission NRF setup

3. Radioscopy and computed tomography with brilliant gamma beams

Industrial gamma radioscopy and computed tomography using laser-Compton scattering (LCS) gamma beams have seen an increased interest lately because of the attractive characteristics of LCS beams as quasi monoenergetic, high intensity and high energy gamma beams [13]. It can be anticipated that these applications will benefit greatly from the unique characteristics of the ELI-NP LCS gamma beam [2]. The high-energy, high-intensity and narrow bandwidth (<0.5%) features of the gamma beam will allow investigation of large industrial objects with unprecedented resolution and practically free of beam hardening.

The setup that will be implemented at ELI-NP will specialize in non-destructive testing and analysis of various size objects by producing 2D transmission images and/or 3D reconstructed tomographs to reveal their internal fine structure and composition. These non-destructive analyses can be used to test the quality of industrial objects, e.g. inspect sinterizations, weldings or cracks in large objects. Two experimental configurations are under preparation: a pencil-beam and a cone-beam configuration. The pencil-beam configuration is sketched in Fig. 2. A high volume detector will absorb almost entirely the transmitted pencil beam while an automated positioning system capable of biaxial translation and rotation will be available for placing objects (up to 150 Kg) with high accuracy in beam. For cone-beam CT the high-volume detector will be replaced with a 2D gamma-ray imaging system.

An estimate on the performance of these setups was obtained using analytical methods and numerical simulations in GEANT4 [14, 15]. We estimated analytically the contrast of a stainless steel (SS) grid (see Fig. 2) against the background provided by homogeneous objects of various composition and in different configurations (see Ref. [2] for more details). For example, table 2 lists the expected thickness of several materials for which a contrast corresponding to 0.7 mm SS can be obtained in a single exposure of one second. Lots of exposures (as exemplified below) are needed to produce a radiograph or a CT image hence exposure times as small as possible are desired. The values listed in the table are produced using a source-object distance of 20 m, a magnification factor of 2, a detector efficiency of one and the gamma source intensity and divergence listed in Table 1. For instance, at 3.5 MeV one can obtain a contrast of 1.9 % in Al objects up to 40 cm in size and the same contrast in 13 cm of Fe block. The thickness doubles and almost triples at 9.87 and 19.5 MeV, respectively, thanks to the improved parameters of the gamma beam at high energy.

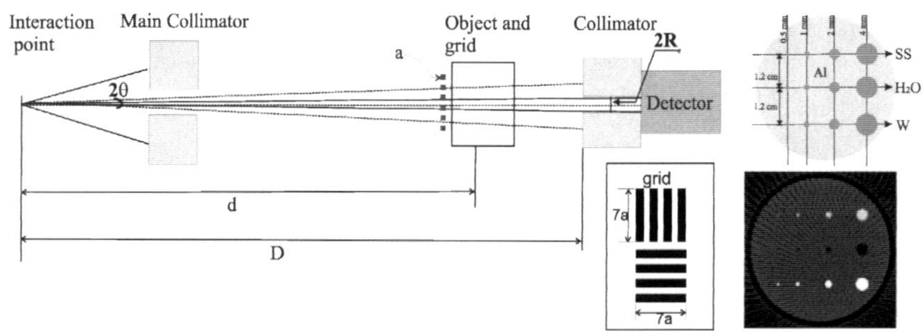

Figure 2. (a) Schematic drawing of the experimental setup and of the grid used to assess the resolution. A 2D CT image of phantom obtained from simulated data is shown on the right.

To test the spatial resolution and the contrast sensitivity of the tomography setup we have simulated the transmission through various objects and grids using Monte Carlo simulations implemented in GEANT4. We confirmed that photons scattered inside the object don't reach the detector and that images are free of beam hardening. Fig. 2 presents an example of a simulated 2D cross-sectional CT image of a phantom using the phantom composition shown in the same figure. The 2D cross-sectional image was obtained using filtered back-projection algorithm from 60 projections (3 degrees projection angle) with 109 views per projection (0.5 mm step). Each exposure was produced using a beam of 1×10^6 photons/s of 3.5 MeV$\pm$17 keV and 0.5 mm beam spot [16]. Sub-millimeter (0.5 mm) features are best visible for high-density materials (higher contrast). In order to discriminate lower contrast parts (thin and low density objects) a smaller beam spot is desirable, which implies further collimation. Note that a smaller beam spot ($<$ 0.5 mm) imply greatly reduced beam intensity for any conventional gamma source but not for the future gamma beam at ELI-NP, which is inherently narrow (divergence between 200 μrad and 40 μrad, see Ref. [6]) and the perfect solution for high-resolution tomography.

Table 2. Estimated thickness of given materials that can be scanned with 0.7 mm resolution.

Energy (MeV)	2	3.5	9.87	19.5
Al (cm)	30	41.2	88.9	104.9
Fe (cm)	10.4	13.5	23.7	24.3
H_2O (cm)	70.8	100.5	250.1	337.05
Concrete (cm)	33.8	47	106.8	131.6
Photons/macropulse	1.9×10^3	2.9×10^3	3.9×10^3	3.4×10^3

While the performance of the radioscopy and computed tomography setup rely on the unique characteristics of the gamma beam it depends greatly on the detector performance. Table 2 lists also the number of photons per macropulse that a detector experiences for applications that require high resolution and short exposure time for a pencil-beam setup. These values represent the number of photons that arrive at the detector in 32 bunches of picosecond duration separated by 16 ns. To cope with such high instant rates the detector has to be fast and radiation hard. The available scintillator detectors in use nowadays for tomography experiments have to be used with caution at ELI-NP because accidental absorption of full uncollimated beam can lead to radiation damage in the scintillator and high peak currents that will exceed the maximum values accepted for the photomultipliers (PMT). A search for alternative detection schemes for pencil-beam tomography is underway.

4. Scintillator based gamma beam diagnostics instruments

The gamma beam produced at ELI-NP has unprecedented characteristics which call for constant monitoring and adjustment of its parameters to ensure proper operation. Most of the nuclear physics experiments and the applications described above require continuous and precise knowledge of the operational conditions, such as the beam energy and energy spread, beam intensity, the degree of polarization, the temporal structure and the spatial position.

Monitoring the parameters of the gamma-ray beam demands instruments capable of operating in radiation harsh environments, with fast response and excellent energy resolution. Several instruments based on scintillator materials have been proposed for this task because of their fast response time. An intensity and polarization monitor based on Li-glass and NE213-type detectors is proposed to measure the outgoing neutrons from the d(γ,n)p reaction [17, 18]. The gamma beam energy spread above 10 MeV will be monitored using a large volume $LaBr_3$ detector with a NaI(Tl) anti-Compton shield placed in the attenuated beam [17]. Several additional instruments based on $LaBr_3$ detectors are using Compton scattering and NRF for measuring the time structure, the intensity, and the polarization of the beam [17]. A thin LYSO scintillator sheet is foreseen to be placed in the gamma beam and in conjunction with a CCD camera would give insight into the spatial position of the beam and allow quick experimental instrument alignment.

The measurement of the beam intensity and polarization at energies above 3 MeV will take advantage of the large cross section and polarization asymmetry inherent in the d(γ,n)p reaction. The number of emitted neutrons peaks in the plane of beam polarization, and relatively few neutrons are emitted perpendicular to the plane of beam polarization. The neutron detectors should be oriented as shown in Fig. 3. The detector located at 45° with respect to the plane of beam polarization will be used to measure any misalignment in the plane of beam polarization or any component of circular polarization. This technique has been tested successfully at the HIγS facility [18].

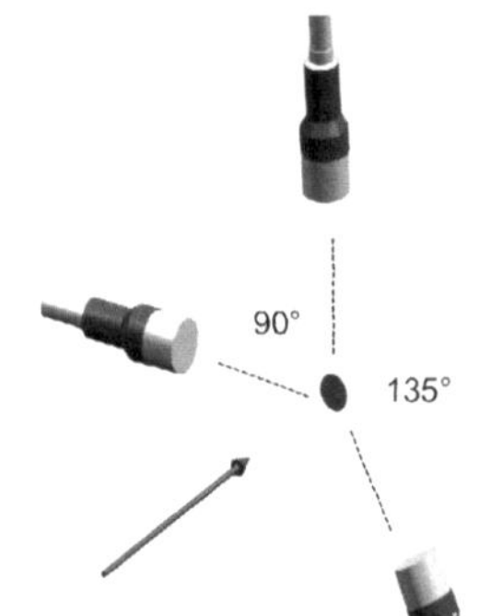

Figure 3. A schematic view of the relative position of the detectors for the polarization and intensity monitoring using either the NRF and the d(γ,n)p reaction. The beam is indicated by the red arrow, and the target is shown as a yellow disk.

The choice of neutron detector for the d(γ,n)p setup depends on the beam energy. For beam energies below 4 MeV, Li-glass detectors will be used to detect the outgoing neutrons. These detectors can count with greater efficiency the 200–900 keV neutrons and are not subject to the same threshold effects as the liquid scintillator neutron detectors. For beam energies above 4 MeV, liquid scintillator neutron detectors based on EJ-309 scintillation material will be used. The advantages of using EJ-309 are its detection efficiency, which has been thoroughly studied, and its good pulse shape discrimination (PSD) properties. Moreover, the EJ309 detectors may be able to resolve nearby pulses since their short component decay time is around 4 ns. A conservative count rate estimate of 2 counts per macropulse, or 200 Hz, would allow for a measurement of the beam polarization to a statistical precision of 1% within a few minutes.

For time structure measurement, $LaBr_3(Ce)$, with its 16 ns 1/e decay time, is indeed a very promising option for resolving individual pulses. Unfortunately, the rise and fall times increase with increasing crystal size, an effect arising from the increasing path length of scintillation photons in larger crystals. This effect is amplified by the difference in refractive index between $LaBr_3$ and the boro-silicate glass used in the PMT windows, so that scintillation photons undergo many reflections before absorption at the photocathode. Tests at HIγS facility with a 16 ns time structure similar to the one developed at ELI-NP, demonstrate that a one inch detector is capable of reaching the anticipated decay time of 16 ns and the resulting spectra can be deconvoluted.

5. Conclusions

The gamma beam system at ELI-NP will deliver a very intense, quasi-monoenergetic and energy tunable gamma beam. This will be a perfect solution for industrial tomography applications as corroborated by our estimations based on analytical calculations and numerical simulations. The non-destructive assay of bulk materials based on NRF of can be successfully applied at ELI-NP thanks to the high-intensity gamma source and the high efficiency detector array. Scintillator based instruments are proposed for monitoring the characteristics of the gamma beam.

Acknowledgments

The author(s) would like to acknowledge the help received from the engineering bureau, especially from Cristian Petcu and Emil Udup. The ELI-NP project is co-financed by the Romanian Government and European Union through the European Regional Development Fund.

References

[1] Zamfir N V 2016, *EPJ Web of Conferences* **117** 01001
[2] Suliman G, Iancu V, Ur C A, Iovea M, Daito I, Ohgaki H 2016 *Rom. Rep. Phys* **68** S799
[3] Bobeica M *et al.* 2016 *Rom. Rep. Phys* **68** S847
[4] Asavei T *et al.* 2016 *Rom. Rep. Phys* **68** S275
[5] Djourelov N, Hugenschmidt C, Balascuta S, Leca V, Oprisa A, Piochacz C, Teodorescu C and Ur C A 2016 *Rom. Rep. Phys* **68** S735
[6] Adriani O *et al.*, 2014 Technical Design Report EuroGammaS proposal for the ELI-NP Gamma beam System, *Preprint* arXiv:1407.3669
[7] Ludewigt B A, Quiter B J and Ambers S D 2011 *Nuclear resonance fluorescence for safeguards applications* (DOE report, url: http://dx.doi.org/10.2172/1022713.)
[8] Bertozzi W and Ledoux R J 2005 *Nucl. Instr. and Meth. Phys. Res.* B **241**, 820
[9] Toyokawa H et al. 2011 *Jap. J. of Appl. Phys.* **50** 100209
[10] Lakshmanan M N, Harrawood B P, Agasthya G A, Kapadia A J 2014 *IEEE Trans. On Med. Imag.* **33** 546
[11] Ur C A *et al.* 2016 *Rom. Rep. Phys* **68** S483
[12] Daito I, Ohgaki H, Suliman G, Iancu V, Ur C A and Iovea M 2016 *Energy Procedia* **89** 389
[13] Toyokawa H, Ohgaki H, Mikado T, and Yamada K 2002 *Rev. of Sci. Instrum.* **73** 3358
[14] Agostinelli S *et al.* 2003 *J. Nucl. Instr. Meth. Phys. Res.* A **506** 250
[15] Allison J*et al.*, 2006 *IEEE Trans. Nucl. Sci.* **53** 270
[16] Suliman G, Iancu V, Ur C A, Iovea M, Daito I, Ohgaki 2016 *Int. J. of Modern Physics: Conference Series* **44** 1660216
[17] Weller H R *et al.* 2016 *Rom. Rep. in Phys.* **68** S447
[18] Matei C, Mueller J M, Sikora M H, Suliman G, Ur C A and Weller H R 2016 *J. of Instrum* **11** P05025

Journal of Physics: Conference Series **763** (2016) 012004 doi:10.1088/1742-6596/763/1/012004

Applications of $LaBr_3(Ce)$ Gamma-ray Spectrometer Arrays for Nuclear Spectroscopy and Radionuclide Assay

PH Regan[1,2,+], R Shearman[1,2], T Daniel[1], G Lorusso[1,2], SM Collins[2], SM Judge[1,2], SJ Bell[2], AK Pearce[2], LA Gurgi[1], M Rudigier[1], Zs Podolyák[1], N Mărginean[3], R Mărginean[3], S. Kisyov[3]

[1]Department of Physics, University of Surrey, Guildford, GU2 7XH, UK
[2]AIR Division, National Physical Laboratory, Teddington, TW11 0LW, UK
[3]Horia Hulubei National Institute of Physics and Nuclear Engineering, Maguerele, R-077125, Bucharest, Romania

[+]Corresponding author: p.regan@surrey.ac.uk ; paddy.regan@npl.co.uk

Abstract. An overview of the use of discrete energy gamma-ray detectors based on cerium-doped $LaBr_3$ scintillators for use in nuclear spectroscopy is presented. This review includes recent applications of such detectors in mixed, 'hybrid' gamma-ray coincidence detection arrays such ROSPHERE at IFIN-HH, Bucharest; EXILL+FATIMA at ILL Grenoble, France; GAMMASPHERE+FATIMA at Argonne National Laboratory, USA; FATIMA + EURICA, at RIKEN, Japan; and the National Nuclear Array (NANA) at the UK's National Physical Laboratory. This conference paper highlights the capabilities and limitations of using these sub-nanosecond 'fast-timing', medium-resolution gamma-ray detectors for both nuclear structure research and radionuclide standardisation. Potential future application of such coincidence scintillator arrays in measurements of civilian nuclear fuel waste evaluation and assay is demonstrated using coincidence spectroscopy of a mixed $^{134,7}Cs$ source.

1. Introduction: The Use of $LaBr_3(Ce)$ Detectors for Nuclear Gamma-ray Measurements.

The science of nuclear spectrometry requires careful measurement of the characteristic decay properties of a range of radionuclides. These can have applications in nuclear waste disposal, radiopharmaceutical standardisations and also in fundamental studies of the internal structure of atomic nuclei. The characteristic, discrete-energy gamma-ray emissions from excited states populated following radioactive decay provide unique signatures which can be used both for identification and quantification of different radionuclides species in mixed-material sources. The workhorse detectors for such studies over the last three decades have been hyper-pure germanium (HPGe) semiconductor detectors which have excellent gamma-ray energy resolution (<2 keV FWHM at 1.33 MeV full energy peak) but have to be operated at liquid nitrogen temperature in order to overcome thermal excitation noise effects.

Over the course of the last decade there has been a significant development in nuclear instrumentation based on detectors made from halide scintillation materials for use in gamma-ray spectroscopy and spectrometry. In particular, the inorganic scintillator Lanthanum Tribromide with Cerium doping ($LaBr_3(Ce)$) is becoming a mainstream tool in nuclear spectroscopy research experiments [2-20]. This material has a significantly poorer energy resolution compared to HPGe (~40 keV FWHM at 1.33 MeV [7]) but is much better than other scintillator materials for gamma-ray measurement such as NaI(Tl) or CsI [1]. It also has some operational advantages compared to HPGe in that its intrinsic timing resolution is sub-nanosecond across a wide energy range (from ~100 keV up to more than 4 MeV) reduces random events in energy coincidence analyses where two or more discrete energy gamma rays are emitted in a sequential cascade following a radioactive decay. It can also provide precision determinations of electromagnetic decay lifetimes of intrinsic, excited nuclear states, thereby providing tests of physical models of nuclear structure.

This ANSRI conference paper reviews the construction and application of a range of gamma-ray spectrometer arrays worldwide which utilise contemporary $LaBr_3(Ce)$ detector technology and provide examples of analysis for nuclear waste assay and nuclear structure physics measurements.

2. The National Nuclear Array – NANA

The National Nuclear Array (NANA) has been commissioned at the UK National Physical Laboratory [2-4]. This array comprises 12 $LaBr_3(Ce)$ detectors, each of which is a cylinder of length 51 mm and circumference 38 mm. The current array design consists of a central ring of 8 symmetrically-spaced $LaBr_3(Ce)$ detector modules with an additional ring of 4 detectors off-set at backward angles. These are the same detector modules and geometries as used in the DESPEC-FATIMA gamma-ray spectrometer developed for use in nuclear structure decay experiments at Radioactive Ion Beam research facilities such as at GSI/FAIR and RIBF-RIKEN [5-8].

Figure 1 shows photographs of the NANA in-situ at the National Physical Laboratory. The $LaBr_3(Ce)$ detectors are coupled to Hammatsu R9779 photomultiplier tubes (PMTs) were are chosen to maximise the timing capabilities of these detectors. The energy signals from the PMTs are taken directly into CAEN V1751, 1 GHz digitisers which provide independent but synchronised time-stamped, energy signals for each input channel. The digitised nanosecond timing allows event-by-event coincidence information between gamma rays emitted from a discrete cascade to be temporally correlated; in turn, logical gating conditions can be set on particular gamma-ray energies which define a unique energy-level decay sequence characteristic for a particular radionuclide species. This coincidence gating property can be applied event-by-event in off-line software data analysis and allows $LaBr_3(Ce)$-based detection arrays such as NANA the potential to identify relatively low-activity concentrations. This can be particularly useful when analysing mixed-waste source matrices which may emit a large number of different energy gamma rays resulting in what, at first instance, leads to a rather messy and hard to interpret 'singles' spectrum.

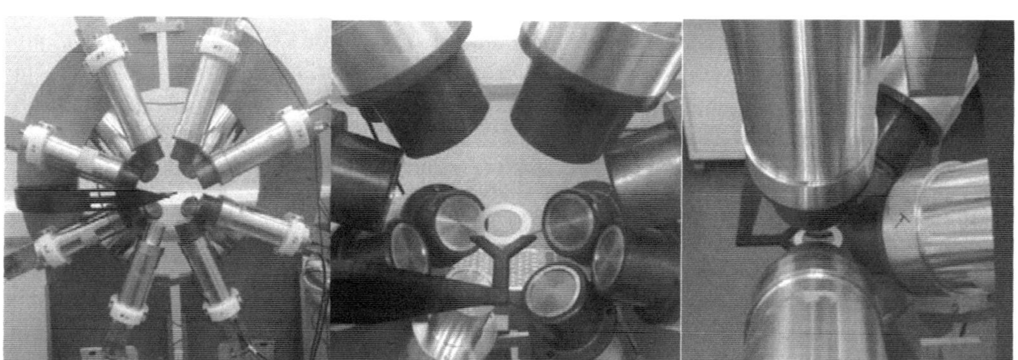

Figure 1 The NANA gamma-ray spectrometer at the UK National Physical Laboratory, comprising twelve individual $LaBr_3(Ce)$ gamma-ray detector modules.

2.1. Example of use of NANA for Radionuclide Assay: ^{134}Cs/^{137}Cs Identification and Separation.

The NANA allows the use of coincidence gamma-ray spectroscopy in the evaluation of radionuclide materials. As an example, we have taken data using six independent, modules within the NANA array from gamma rays emitted from a well characterised mixed ^{134}Cs/^{137}Cs source. Both of these radioistopes would be expected to be present in nuclear waste fuel; (a) ^{137}Cs, which has a half-life of 30 years [22], as a direct fission fragment daughter and (b) ^{134}Cs ($T_{1/2}$=2.0652 years [23-24]) which is formed in-situ in nuclear reactor fuels following thermal neutron-capture on the ^{133}Cs fission daughter residue which builds up during normal reactor operation.

The gamma-ray decays following the radioactive decay for these two isotopes are notably different with ^{137}Cs decaying primarily to a single excited, spin/parity I^{π}=11/2$^-$ metastable state ($T_{1/2}$=2.55 mins) which then emits a single 662 keV gamma-ray to the I^{π}=3/2$^+$ ground state of the ^{137}Ba daughter radionuclide. By contrast, the spin/parity I^{π}=4$^-$ ground state of ^{134}Cs decays both by electron capture to ^{134}Xe and also by β^- decay to excited states in ^{134}Ba [23-24]. The left hand side of Figure 2 shows a simplified partial level schemes for the states populated in ^{134}Ba following the radioactive decay of ^{134}Cs. Note that there is a range of multi-gamma-ray cascades which feed the 605 keV yrast I^{π}=2$^+$ state in the ^{134}Ba daughter. The right hand side of Figure 2 shows the summed singles spectra for 6 detectors of the NANA array from the mixed ^{134,7}Cs source. The raw, energy-calibrated spectra show the well-understood signatures of arising from the internal activity within the $LaBr_3$(Ce) detectors from (i) the decay of the ^{138}La primordial isotope [25] and (ii) a cascade of alpha emitters from the members of the actinium natural decay series (4n+3) chain headed by ^{227}Ac which are intrinsic to the detector material. A normalized fraction of the 'sourceless' spectrum (i.e., a spectrum taken from detector signals measured with no source present in the NANA source position) has been subtracted and this clearly shows and is able separate discrete lines from the decay of ^{137}Cs (662 keV) and ^{134}Cs (569, 605, 796, 1038, 1168 and 1365 keV).

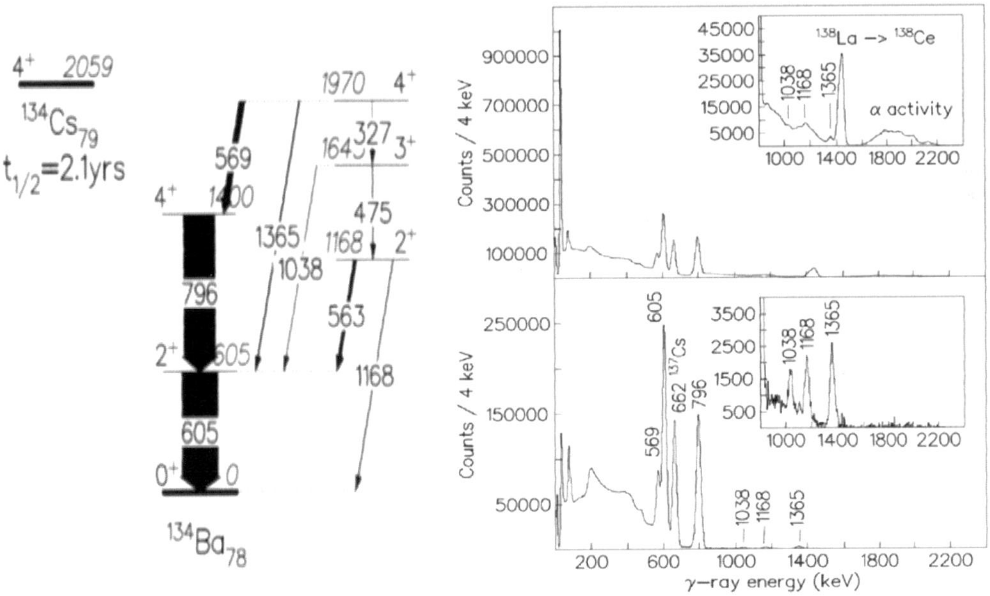

Figure 2 (Left) Partial decay scheme of states populated in ^{134}Ba following the radioactive decay of the nuclear waste product ^{134}Cs [23,24]. (Right) summed singles (upper panel) and background-subtracted (lower panel) $LaBr_3$(Ce) spectra using NANA for a mixed ^{134,137}Cs source. Note the clear identification of the weak transitions above 1 MeV in the background subtracted spectrum.

The selection power of NANA in using the coincidence method to isolate gamma-ray decay cascades associated with particular radionuclide decays is demonstrated in Figure 3. These are the same data as shown in Figure 2 but with an offline software-sorting condition applied that at an energy signal should be registered in at least two separate detectors of the NANA array within a time window of ± 5 ns. For these data, six of the full complement twelve NANA detector modules were operational. The left hand side of Figure 3 shows the two-dimensional gamma-ray-energy coincidence matrix for these data. The diagonal, negative gradient stripes in this matrix correspond to Compton scattered events where the sum of energies in the two coincident detectors sum to a single (Compton scattered) discrete energy. Notable lines of this nature include the 1435.8 keV transition emitted from the internal ^{138}La activity in the detector material [25]. By contrast, the main two-dimensional peaks within this energy matrix correspond to the full-energy peak coincidence events from the decay of ^{134}Cs into excited states in ^{134}Ba, for example the coincidence pairs at {605,796} and {605;563/569} [23,24]. The right hand side of Figure 3 shows a total projection (i.e. sum) of all coincidence with ≥ 2 independent $LaBr_3$(Ce) events measured within a ±5 ns coincidence window. Note that there is no lines present in this spectrum associated with the 662 keV transition from the ^{137}Cs decay, even though this line is clearly observed in the singles spectra for the same data shown on the left hand-side of Figure 2. Furthermore, when off-line coincidence conditions are set whereby a discrete full-energy peak is measured in one of the two coincidence detectors, the projections can be used to demonstrate and isolate different decay paths within the ^{134}Ba nucleus. For example, when a coincidence is required with the 1365 keV transition, only the 605 keV $2^+ \rightarrow 0^+$ transition from the decay of first excited state in ^{134}Ba is evident in the spectrum, with the 796 keV and other transitions absent.

The use of $LaBr_3$(Ce) detectors has been mooted recently for application in nuclear waste fuel assay [26]; the additional parameter dimension which is allowed by discrete gamma-ray coincidence analysis could allow an improvement of this technique for more detailed assay of more weakly populated fission residues if their decay incorporates a coincidence decay sequence such as the ^{134}Cs example shown in the current work.

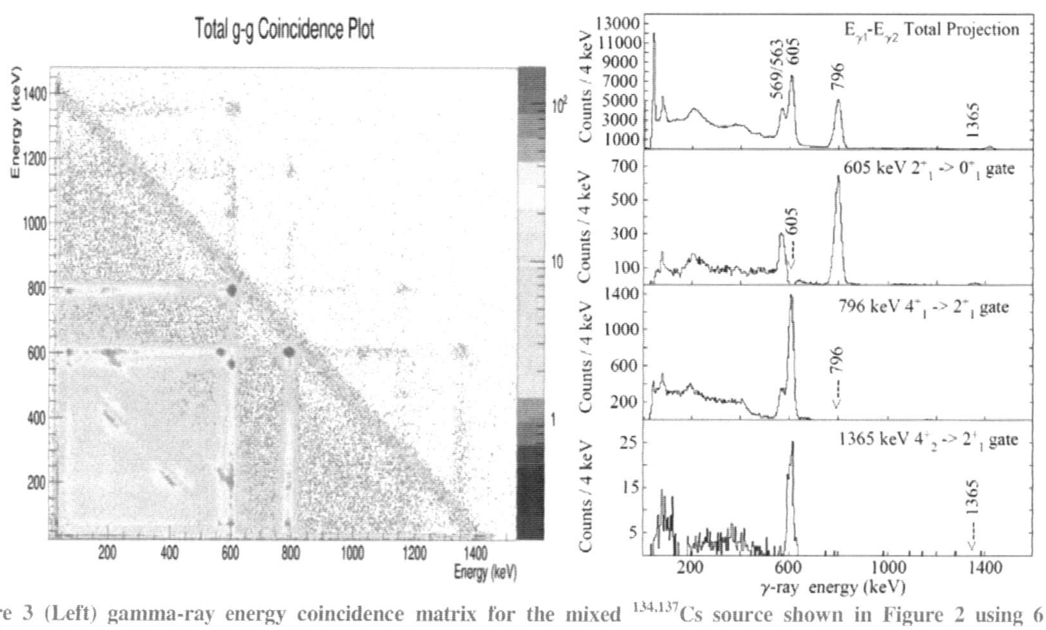

Figure 3 (Left) gamma-ray energy coincidence matrix for the mixed ^{134,137}Cs source shown in Figure 2 using 6 individual detectors in the NANA array; (Right) Total Projection (upper spectrum) and Compton-background subtracted gates on the matrix showing coincidence transition in ^{134}Ba populated following the decay of ^{134}Cs.

3. Hybrid-$LaBr_3$(Ce)-HPGe Gamma-ray Coincidence Arrays in Nuclear Structure Research:

3.1. The DESPEC-FATIMA Array for FAIR and Related sub-arrays.

A number of coincidence gamma-ray spectrometer arrays have been constructed and commissioned worldwide to exploit the timing characteristics of $LaBr_3$(Ce) detectors in nuclear spectroscopy with focus on the measurement of absolute electromagnetic transition rates associated with decays from excited, particle bound nuclear states. The detectors in the NANA are also used in the DESPEC Fast-TIMing Array (FATIMA). This array consists of three concentric rings of twelve $LaBr_3$(Ce) detectors (36 detectors in total) and has been designed for gamma-ray spectroscopic measurement following decays in nuclei with very exotic proton-to-neutron ratios produced at high-energy nuclear projectile fragmentation facilities such as GSI/FAIR and RIBF-RIKEN [5-8].

Some of these $LaBr_3$(Ce) detector modules have been incorporated into mixed, 'hybrid' arrays which combine HPGe detectors for high-resolution gamma-ray selection and $LaBr_3$(Ce) modules for coincident nuclear fast-timing measurements. An example is the EXILL+FATIMA set-up which was used to measure prompt gamma-ray coincidences following the cold-neutron-induced, prompt fission of ^{235}U at ILL Grenoble [11-12]. EXILL+FATIMA comprised eight $LaBr_3$(Ce) detectors used in concert with eight, four-element HPGe Clover detectors in a compact geometry and was been used to study the internal structure of neutron-rich, prompt-fission residues. The system typically used quadruple coincidence events (2 x HPGe and 2 x $LaBr_3$(Ce)) to isolate individual fission fragment cascades from the hundreds of identified prompt species populated in this reaction mechanism. The EXILL-FATIMA set-up has been used to measure nanosecond excited state lifetimes for the excited states of neutron-rich nuclei such as ^{100}Zr [11] and ^{90}Kr [12].

The DESPEC-FATIMA detectors have also been incorporated at the Radioactive Ion Beam Facility (RIBF) at RIKEN, Japan for related nuclear structure studies of some the most neutron-rich nuclear species identified to date. A mixed, hybrid array of 18 DESPEC-FATIMA $LaBr_3$(Ce) detectors were coupled to twelve, seven-element HPGe 'CLUSTER' detectors at the focal plane of the BigRIPS projectile separator at RIKEN. The main application for this array was gated isomer and beta-delayed nuclear spectroscopy in very neutron-rich nuclei which were produced following the high-energy projectile fission of a ^{238}U primary beam [13]. Recent physics highlights using this array include the measurement of the first excited state lifetimes in ^{104,6}Zr which are 8 and 10 neutron heavier respectively than the most neutron-rich stable Zirconium isotope [14]. This new experimental information can be used to determine the low-lying shape of these mesoscopic physical systems and allows a comparison with state of the art nuclear mean-field energy density functional calculations which predict the evolution of nuclear structure in such nuclei of astrophysical interest. Twenty-five of the DESPEC-FATIMA $LaBr_3$(Ce) detectors were coupled together with 55 HPGe Compton suppressed detectors of the GAMMASPHERE array at Argonne National Laboratory in late 2015 for an experimental measurement campaign of prompt and delayed decays from spontaneous fission residues from a sealed ^{252}Cf source [8]. An array of 8 $LaBr_3$(Ce) FATIMA detectors has also been deployed at the focal plane of the RITU gas-filled mass separator at JYFL- Jyvaskyla, Finland for lifetime and nuclear deformation measurements in N~74 isotones following isomeric decay [27].

Experimental plans for the 2016-17 period include combining these FATIMA detector modules with the AGATA gamma-ray tracking spectrometer for a nuclear structure campaign at the GANIL laboratory in Caen, France for studies of neutron-rich nuclear systems using heavy-ion induced deep-inelastic reactions and also with 24 clover HPGE detectors to form the NuBALL hybrid gamma-ray spectrometer to be used at IPN-Orsay, Paris France for fast-neutron induced fission measurements.

Journal of Physics: Conference Series **763** (2016) 012004 doi:10.1088/1742-6596/763/1/012004

3.2. RoSPHERE

The first major hybrid $LaBr_3$(Ce)-HPGe array to be used in continuous mode is the **RO**manian array for **SP**ectroscopy in **HE**avy ion **RE**actions (ROSPHERE) at IFIN-HH, Magurele, Romania [28,29]. Its most recent configuration comprises 25 individual detectors, fourteen 50-60% relative efficiency Compton Suppressed HPGe spectrometers and 11 $LaBr_3$(Ce) modules, with five detectors each set into five concentric rings at angles of 37°, 70°, 90°, 110° and 143° to the beam direction. Figure 4 shows photographs of the ROSPHERE array with the open vacuum target chamber.

Figure 4 Insert of part of the ROSPHERE hybrid gamma-ray at IFIN-HH, Romania consisting of Compton Suppressed HPGe detectors and $LaBr_3$(Ce) for fast-timing measurements [29].

ROSPHERE has been used to study a wide range of nuclear structure phenomena in nuclear created following heavy-ion reactions from beams produced by the 9 MV tandem van de Graaff accelerator at IFIN-HH. Such studies have included evaluation of single-particle transitions around the N=20 [15], N=82 [16], and N=126 [17] magic numbers; measurements of low-lying octupole (E3) excitations in neutron-rich copper nuclei [18,19]; shape evolution from prolate to oblate deformations in tungsten isotopes [20]; and the robustness of the K-quantum number associated with axial symmetry in prolate-deformed Cerium isotopes [21].

Figure 5 shows example spectra from ROSPHERE following an 80 MeV ^{18}O beam impinging on an isotopically enriched self-supported ^{192}Os target. The figure highlights the difference in spectral resolution response between the HPGe and $LaBr_3$(Ce) detectors of ROSPHERE with the superior resolution of the HPGe components clearly evident. The energy resolution of the $LaBr_3$(Ce) detectors is however, in many cases sufficient to isolate specific decay sequences using the gamma-ray energy coincidence method as discussed in Section 2. This is demonstrated in Figure 6 which show background subtracted energy coincidence gates on the decays from the first $I^\pi=2^+$ and $I^\pi=4^+$ states in in ^{192}Os.

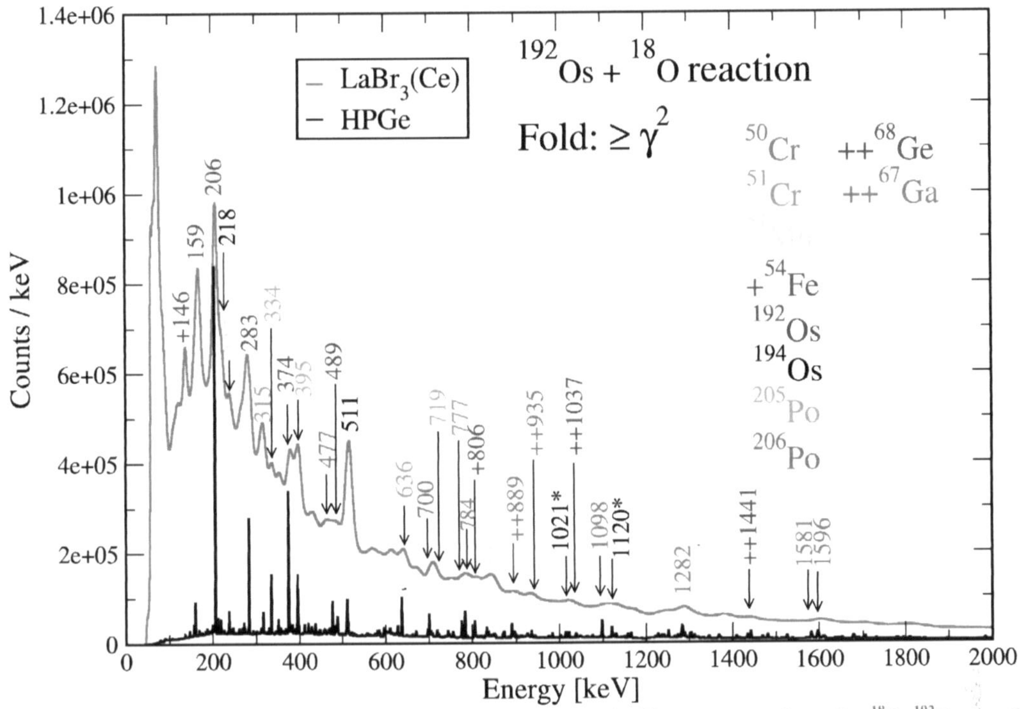

Figure 5 Total projections of the HPGe and $LaBr_3(Ce)$ gamma-ray coincidence spectra from the $^{18}O+^{192}Os$ reaction performed at IFIN-HH using the ROSPHERE spectrometer. Discrete peaks are labelled in the HPGe spectra following reactions on the main target material and also following fusion of the ^{18}O beam with target contaminants associated with ^{28}Si, ^{40}Ca and ^{54}Fe.

Figure 6 shows a time difference spectra obtained by gating on the $LaBr_3(Ce)$ detector full energy peaks at 206 and 374 keV in ROSPHERE. These correspond to the discrete transitions which feed into and out of the $I^{\pi}=2^+$ yrast state in ^{192}Os [30]. The half-life of this excited nuclear state can then be extracted from these data by evaluating the centroid of the displaced time-difference distributions forward and backward gated in time. The lifetime can also be obtained by fitting the time-difference distribution to a function which is a convolution of the Gaussian prompt response function (PRF) for the $LaBr_3(Ce)$ detectors (which is energy dependent) together with a single exponential decay lifetime associated with the decay of the state of interest. Figure 6 shows both of these analysis methods for two sets of gates which feed into and out of the yrast $I^{\pi}=2^+$ state in ^{192}Os and show results for intrinsic state lifetime which are all consistent with the evaluated value for this state of 288(4) ps [30].

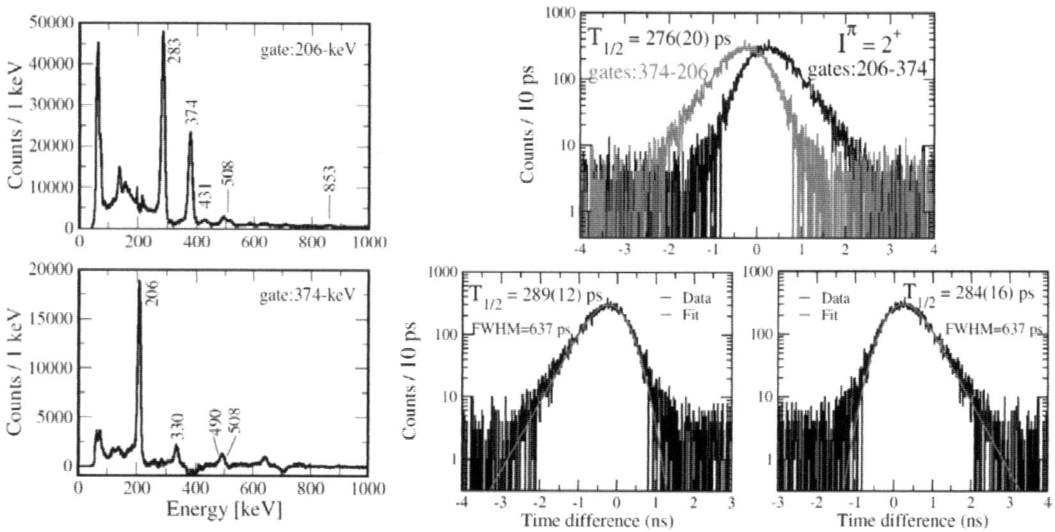

Figure 6 : (Left) Background subtracted gamma-ray coincidence spectra from gated on the $2^+_1 \rightarrow 0^+_1$ (206 keV) and $4^+_1 \rightarrow 2^+_1$ (374 keV) transitions in ^{192}Os [28] populated in unsafe Coulomb excitation following the ^{18}O+^{192}Os reaction. (Right) Background subtracted time-difference spectra gated between the 206 and 374 keV transition which feed into and out of the yrast $I^\pi=2^+$ state in ^{192}Os respectively. The extracted values for the lifetime of this state are consistent with the evaluated value of 288(4) ps [30]. The fit assumes a single exponential decay half-life convoluted with a Gaussian prompt response function with a full-width half-maximum of 637 ps.

4. Summary and Conclusions

The application of $LaBr_3$(Ce) detectors for nuclear spectrometry has been reviewed with emphasis on the potential uses of gamma-ray coincidence measurements for both radioactive waste assay and nuclear structure research. The NANA spectrometer which is housed at the UK's National Physical Laboratory utilised instrumentation and data analysis methodologies developed in the nuclear structure research field and will provide a dedicated national facility for radionuclide assay, radionuclide standardisations and nuclear data measurements in the coming decade.

5. Acknowledgements

The work on NANA has been supported by the UK Department of Business Innovation and Skills through the National Measurement System. PHR, MR and ZP also acknowledge support from the Science and Technology Facilities Council (STFC UK) via grant number ST/L005743/1. RS acknowledges support from an NDA PhD bursary scholarship. The authors are grateful for ongoing collaborations linked to the FATIMA-DESPEC project [31], in particular those with A.M. Bruce, S.Lalkovski, R.Carroll, O. Roberts, C. Nita and D.M. Cullen. The research staff at IFIN-HH, Romania are thanked for their ongoing work with the ROSPHERE array.

References

[1] Bell, S et al., these proceedings.
[2] Lorusso, G., Shearman, R., Regan, P.H., Judge, S.M., Bell, S., Collins, S.M., Larijani, C., Ivanov, P., Jerome, S.M., Keightley, J.D. and Lalkovski, S., 2016. Development of the NPL gamma-ray spectrometer NANA for traceable nuclear decay and structure studies. *Applied Radiation and Isotopes*, *109*, pp.507-511.
[3] Regan, P.H., Shearman, R., Judge, S.M., Lorusso, G., Main, P., Bell, S., Collins, S.M.,

Ivanov, P., Jerome, S.M., Keightley, J.D. and Larijani, C., 2015. Development of NANA: A Fast-Scintillator, Coincidence Gamma-ray Array for Radioactive Source Characterisation and Absolute Activity Measurements at the UK National Physical Laboratory. In *Journal of Physics: Conference Series* (Vol. 620, No. 1, p. 012005). IOP Publishing.

[4] Matta, A., Morfouace, P., De Séréville, N., Flavigny, F., Labiche, M. and Shearman, R., 2016. NPTool: a simulation and analysis framework for low-energy nuclear physics experiments. *Journal of Physics G: Nuclear and Particle Physics*, *43*(4), p.045113.

[5] Regan, P.H., 2015. Precision measurement of sub-nanosecond lifetimes of excited nuclear states using fast-timing coincidences with $LaBr_3$ (Ce) detectors. *Radiation Physics and Chemistry*, *116*, pp.38-42.

[6] Roberts, O.J., Bruce, A.M., Regan, P.H., Podolyák, Z., Townsley, C.M., Smith, J.F., Mulholland, K.F. and Smith, A., 2014. A $LaBr_3$: Ce fast-timing array for DESPEC at FAIR. *Nuclear Instruments and Methods in Physics Research Section A: Accelerators, Spectrometers, Detectors and Associated Equipment*, *748*, pp.91-95.

[7] Regan, P.H., 2012. From RISING to the DESPEC fast-timing project within NUSTAR at FAIR: sub-nanosecond nuclear timing spectroscopy with $LaBr_3$ scintillators. *Applied Radiation and Isotopes*, *70*(7), pp.1125-1130.

[8] Lalkovski, S., Bruce, A.M., Burrows, I., Cullen, D.M., Grant, A., Lazarus, I.H., Podolyák, Z., Pucknell, V.F.E., Regan, P.H., Rudigier, M. and Simpson, J., 2015. Construction of the UK DESPEC Array for Fast-Timing Measurements. *Bulgarian Journal of Physics*, *42*(4).

[9] Regan, P.H., Podolyák, Z., Alharbi, T., Mason, P.J.R., Bruce, A.M., Townsley, C., Roberts, O.J., Mărginean, N., Mărginean, R., Ghită, D. and Mullholland, K., 2013. Precision lifetime measurements using $LaBr_3$ detectors with stable and radioactive beams. In *EPJ Web of Conferences* (Vol. 63, p. 01008). EDP Sciences.

[10] Regan, P.H., 2015, November. Using $LaBr_3$ (Ce) Detectors for Precision Lifetime Measurements of Excited States inInteresting'Nuclei. In *Exotic Nuclei (IASEN-2013)-Proceedings of the First International African Symposium on Exotic Nuclei. Edited by PENIONZHKEVICH YURI ERASTOVICH ET AL. Published by World Scientific Publishing Co. Pte. Ltd., 2015. ISBN# 9789814632041, pp. 579-590* (Vol. 1, pp. 579-590).

[11] Régis, J.M., Simpson, G.S., Blanc, A., de France, G., Jentschel, M., Köster, U., Mutti, P., Paziy, V., Saed-Samii, N., Soldner, T., Ur, C.A. et al.,, 2014. Germanium-gated γ–γ fast timing of excited states in fission fragments using the EXILL&FATIMA spectrometer. *Nuclear Instruments and Methods in Physics Research Section A: Accelerators, Spectrometers, Detectors and Associated Equipment*, *763*, pp.210-220.

[12] Régis, J.M., Jolie, J., Saed-Samii, N., Warr, N., Pfeiffer, M., Blanc, A., Jentschel, M., Köster, U., Mutti, P., Soldner, T. and Simpson, G.S., 2014. B (E 2; $2_1^+ \rightarrow 0_1^+$) value in Kr 90. *Physical Review C*, *90*(6), p.067301.

[13] Patel, Z., Browne, F., Bruce, A.M., Chiga, N., Daido, R., Nishimura, S., Podolyák, Z., Regan, P.H., Roberts, O.J., Sakurai, H. and Söderström, P.A., 2014. Commissioning of a $LaBr_3$(Ce) array with EURICA at RIBF. *RIKEN Accel. Prog. Rep*, *47*, p.13

[14] Browne, F., Bruce, A.M., Sumikama, T., Nishizuka, I., Nishimura, S., Doornenbal, P., Lorusso, G., Söderström, P.A., Watanabe, H., Daido, R., Patel, Z. et al.,, 2015. Lifetime measurements of the first 2^+ states in ^{104,106}Zr: Evolution of ground-state deformations. *Physics Letters B*, *750*, pp.448-452.

[15] Mason, P.J.R., Alharbi, T., Regan, P.H., Mărginean, N., Podolyak, Z., Simpson, E.C., Alkhomashi, N., Bender, P.C., Bowry, M., Bostan, M. and Bucurescu, D., et al., 2012. Half-life of the $I^\pi = 4^-$ intruder state in ^{34}P: M 2 transition strengths approaching the island of inversion. *Physical Review C*, *85*(6), p.064303.

[16] Alharbi, T., Regan, P.H., Mason, P.J.R., Mărginean, N., Podolyák, Z., Bruce, A.M., Simpson, E.C., Algora, A., Alazemi, N., Britton, R. and Bunce, M.R., et al., 2013. Electromagnetic

transition rates in the N= 80 nucleus ^{138}Ce. *Physical Review C, 87*(1), p.014323.

[17] Roberts, O.J., Niţă, C.R., Bruce, A.M., Mărginean, N., Bucurescu, D., Deleanu, D., Filipescu, D., Florea, N.M., Gheorghe, I., Ghiţă, D. and Glodariu, T., et al., 2016. E3 and M2 transition strengths in ^{209}Bi *Physical Review C, 93*(1), p.014309.

[18] Bocchi, G., Leoni, S., Bottoni, S., Benzoni, G., Bracco, A., Bortignon, P.F., Colo, G., Belvito, B., Niţă, C.R., Marginean, N. and Filipescu, D. et al.,, 2014. Probing particle-phonon-coupled states in the neutron-rich nucleus ^{65}Cu by lifetime measurements with fast-timing techniques. *Physical Review C, 89*(5), p.054302.

[19] Nita, C.R., Bucurescu, D., Marginean, N., Avrigeanu, M., Bocchi, G., Bottoni, S., Bracco, A., Bruce, A.M., Cata-Danil, G., Colo, G. and Deleanu, D. et al.,, 2014. Fast-timing lifetime measurements of excited states in ^{67}Cu. *Physical Review C, 89*, p.064314.

[20] Mason, P.J.R., Podolyák, Z., Mărginean, N., Regan, P.H., Stevenson, P.D., Werner, V., Alexander, T., Algora, A., Alharbi, T., Bowry, M. and Britton, R. et al.,, 2013. Half-life of the yrast 2^+ state in ^{188}W: Evolution of deformation and collectivity in neutron-rich tungsten isotopes. *Physical Review C, 88*(4), p.044301.

[21] Alharbi, T., Regan, P.H., Mărginean, N., Podolyák, Z., Roberts, O.J., Bruce, A.M., Alkhomashi, N., Britton, R., Bucurescu, D., Deleanu, D. and Filipescu, D. et al., , 2015. Lifetime of the yrast $I^\pi= 5^-$ state and E 1 hindrance in the transitional nucleus ^{136}Ce. *Physical Review C, 91*(2), p.027302.

[22] Browne, E. and Tuli, J.K., 2007. Nuclear data sheets for A= 137. *Nuclear Data Sheets, 108*(10), pp.2173-2318.

[23] Sonzogni, A.A., 2004. Nuclear data sheets for A= 134. *Nuclear Data Sheets, 103*(1), pp.1-182.

[24] Wang, G., Alburger, D.E. and Warburton, E.K., 1987. Precision energy and intensity measurements for γ transitions from ^{134}Cs decay. *Nuclear Instruments and Methods in Physics Research Section A: Accelerators, Spectrometers, Detectors and Associated Equipment, 260*(2-3), pp.413-417.

[25] Sonzogni, A.A., 2003. Nuclear data sheets for A= 138. *Nuclear Data Sheets, 98*(3), pp.515-664.

[26] Bender, S., Heidrich, B. and Ünlü, K., 2015. Compton suppressed $LaBr_3$ detection system for use in nondestructive spent fuel assay. *Nuclear Instruments and Methods in Physics Research Section A: Accelerators, Spectrometers, Detectors and Associated Equipment, 784*, pp.474-481.

[27] Mallaburn, M.J., Cullen, D.M., et al., A fast-timing correctin method for spatially separated recoil implants at a spectrometer focal plane, to be submitted to *Nuclear Instruments and Methods in Physics Research* (2016).

[28] Mărginean, N., Balabanski, D.L., Bucurescu, D., Lalkovski, S., Atanasova, L., Căta-Danil, G., Căta-Danil, I., Daugas, J.M., Deleanu, D., Detistov, P. and Deyanova, G., 2010. In-beam measurements of sub-nanosecond nuclear lifetimes with a mixed array of HPGe and $LaBr_3$: Ce detectors. *The European Physical Journal* A, 46(3), pp.329-336.

[29] Bucurescu, D., Căta-Danil, I., Ciocan, G., Costache, C., Deleanu, D., Dima, R., Filipescu, D., Florea, N., Ghiţă, D.G., Glodariu, T. and Ivaşcu, M., 2016. The ROSPHERE γ-ray spectroscopy array. *Nuclear Instruments and Methods in Physics Research Section A: Accelerators, Spectrometers, Detectors and Associated Equipment, 837*, pp.1-10.

[30] Baglin, C.M., 2012. Nuclear data sheets for A= 192. *Nuclear Data Sheets, 113*(8), pp.1871-2111.

[31] Podolyák, Z., 2014. Studies of exotic nuclei with advanced radiation detectors. *Radiation Physics and Chemistry, 95*, pp.14-18.

Applications of Novel Scintillators for Research and Industry 2016 (ANSRI 2016) IOP Publishing
Journal of Physics: Conference Series **763** (2016) 012005 doi:10.1088/1742-6596/763/1/012005

Gamma Efficiency Simulations towards Coincidence Measurements for Fusion Cross Sections

M. Heine[1], S. Courtin[1,2], G. Fruet[1], D.G. Jenkins[3], D. Montanari[1,2], L. Morris[3], P.H. Regan[4,5], M. Rudigier[4], D. Symochko[6]

[1] IPHC-CNRS/Université de Strasbourg, 67037 Strasbourg, (France)
[2] USIAS/Université de Strasbourg, 67083 Strasbourg, (France)
[3] University of York, YO10 5DD York (UK)
[4] University of Surrey, GU2 7XH Guildford (UK)
[5] National Physical Laboratory, TW11 0LW Teddington, Middlesex (UK)
[6] IKP/TU Darmstadt, 64289 Darmstadt, (Germany)

E-mail: marcel.heine@iphc.cnrs.fr

Abstract. With the experimental station STELLA (STELlar LAboratory) we will measure fusion cross sections of astrophysical relevance making use of the coincident detection of charged particles and gamma rays for background reduction. For the measurement of gamma rays from the de-excitation of fusion products a compact array of 36 UK FATIMA $LaBr_3$ detectors is designed based on efficiency studies with Geant4. The photo peak efficiency in the region of interest compares to other gamma detection systems used in this field.

The features of the internal decay of ^{138}La is used in a background study to obtain an online calibration of the gamma detectors. Background data are fit to the Monte Carlo model of the self activity assuming crude exponential behavior of external background. Accuracy in the region of interest is of the order of some keV in this first study.

1. Introduction

The measurement of sub-barrier light heavy-ion fusion cross sections can yield both insights into nuclear cluster effects [1] and the *S*-factor at energies of astrophysical interest. In particular, ^{12}C+^{12}C was identified as a key reaction on the production route of heavier elements in massive stars during the carbon burning phase [2], in type Ia supernovae and in superbursts from accreting neutron stars. Though dedicated measurements of sub-barrier ^{12}C+^{12}C cross sections were carried out since the late 1960's, no comprehensive experimental data around the energy regime of astrophysical interest are available at the time being. Hence, the input rates to the nucleosynthesis calculations are uncertain by up to two orders of magnitude for ^{12}C fusion [3].

Since deep sub-barrier fusion reactions are strongly hindered by Coulomb repulsion, the experimental determination of these cross sections, that are expected to be as small as nano barn, is highly challenging. In this configuration the reaction channels are orders of magnitude smaller than background contributions [5] and additional experimental effort is made to determine the fusion cross sections with sufficient accuracy. In the STELLA project described in detail in [4] the determination of such cross sections is targeted with coincidence measurements using the so called gamma-particle-technique [5] based on the time correlated determination of charged particle and gamma energies as well as the measurement of their angular distributions. In addition, the experimental station will be equipped with a rotating target mechanism for sufficient heat dissipation to sustain ^{12}C beam intensities of up to 10μA

For sub-barrier fusion of the ^{12}C+^{12}C system the reaction channels, that include light charged particles, are

$$^{12}\mathrm{C} + ^{12}\mathrm{C} \rightarrow ^{24}\mathrm{Mg}^* \rightarrow ^{23}\mathrm{Na} + \mathrm{p}, \tag{1}$$

$$^{24}\mathrm{Mg}^* \rightarrow ^{20}\mathrm{Ne} + \alpha, \tag{2}$$

with $Q_{\mathrm{p}} = 2.24$ MeV and $Q_\alpha = 4.62$ MeV. The reaction channel with a ^{23}Mg fragment and neutron emission cannot be accessed for relative energies below $Q_{\mathrm{n}} = -2.6$ MeV. The population of the final state of ^{24}Mg with $Q_\gamma = 13.93$ MeV *via* gamma emission is weak by phase space arguments and indeed the yield was found to be negligible in previous studies at higher relative energies [6]. It was furthermore shown that the major deexcitation radiation of the reactions (1) and (2) cascades through the first excited states of the fragment nuclei [7], and hence the detection of characteristic gamma lines at 440 keV and 1634 keV for the identification of the proton and alpha channel, respectively, is targeted in the first instance.

In section 2 the Geant4 simulations for the optimization of the gamma efficiency are addressed and in section 3 a model for the simulation of the self-activity as well as the fit to experimental data are introduced. The contribution is summarized in section 4.

2. Gamma Efficiency Simulations

In order to obtain the highest photo peak efficiency in the gamma detection, comprehensive simulations are carried out for the response to isotropic gamma emission at the target position. For the arrangement of the UK FATIMA $LaBr_3$ crystals [8, 9], detector configurations in a spherical setup, a cylindrical setup where all detectors are facing the beam line, and a simple pile along the direction of the beam are studied. The detectors are stapled starting from 2.5 cm below the equator of the dome of the target chamber in the horizontal plane containing the beam line and the target (see figure 2 in [4]). For efficiency optimization the closest configuration of detectors with spacing smaller 0.2 cm is set. The positions in the lowest line or ring define the entire structure, because going upwards detectors are placed in the gaps of the lower ones. In general, the geometrical acceptance of the single detectors drops by up to 60% towards higher positions with respect to the beam line, because of the target location 2.5 cm below the center of the dome. In this study ten or twelve detectors in the lowest line or ring are considered in order to take shielding effects as well as clustering of energy depositions into account. In table 1 the acceptance of the above mentioned stapling schemes in percent are comprised. Note that

Table 1. Geometrical acceptance in percent of detector alignments in a spherical, cylindrical and wall-like (pile) configuration with ten or twelve detectors in the lowest line or ring.

	Acceptance [%]		
# Det.	Sphere	Cylinder	Pile
10	23.3	23.1	20.6
12	24.4	23.2	21.2

in the spherical setup only 33 and 35 detectors can be placed in the stapling starting with ten or twelve low detectors, respectively. As can be see from table 1, the geometrical acceptance of the pile is significantly lower such that this alignment is ruled out.

The arrangement of ten detectors in the lowest line or ring is less sensitive to additional material like connectors that will be inserted in the beam line and hence this configuration is favored over the configuration with twelve detectors in the lowest segment. The difference of the geometrical acceptance then is negligible (see table 1) and the so called cylindrical setup is constructed. As can be seen in figure 2 in [4], the distance of every second line is optimized with respect to the beam line to cover the upper hemisphere of the experimental station in the most efficient way.

The distance of all pairs of crystals is analyzed to obtain a measure on the geometry independently from the simulation. The parameters of these distributions reflecting the compactness of the detector placement are shown in table 2, where the mean values and the

Table 2. Parameters of the distributions of the pair-wise distance between crystals the spherical as well as cylindrical $LaBr_3$ placement with ten or twelve detectors in the lowest ring or line.

Setup	# Det	Mean [cm]	RMS [cm]
sphere	10	18.3	6.6
sphere	12	18.6	6.9
cylinder	10	19.5	7.2
cylinder	12	20.4	7.8

Root Mean Squares (RMS) in cm are listed. As can be seen, the trend found in the simulation is reproduced and in general the spherical configurations are more compact. Note that all distances are given the same weight. Greater distances in the cylindrical setup can be attributed to detectors with lower geometrical acceptance, *i.e.* lower contribution to (and weight in) the geometrical acceptance in table 1. Hence, the cylindrical contributions appear slightly more disadvantageous in the picture of the pair-wise distances.

For the physics interpretation of experimental data here the photo peak efficiency is of interest. It expresses the percentage of gamma rays depositing their entire energy in the detector volume and is shown in figure 1 in comparison to the experimental performance of the 4π germanium

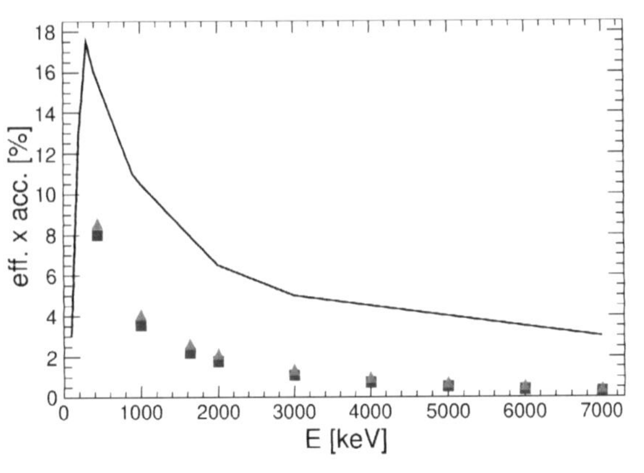

Figure 1. [color online] Comparison of the photo peak response to isotropic gamma emission from the target position with different addback situations (red triangles and blue squares: details in the text) in comparison to the GAMMASPHERE array [10] (black line).

detector array GAMMASPHERE [10, 11] (black line). The current simulation is run with an empty target chamber, since the effect of absorption due to detector material and interior supporting structures is expected to be as low as a few percent of the total photo peak response. All relevant elements of the beam line and required connectors as well as supporting structure in this area are included.

Simulated data shown in figure 1 are analyzed summing up all energy depositions to obtain the total photo peak efficiency (red triangles) and on the other hand summing up the photo peak efficiency of each detector (blue squares). This approach is expected to benchmark the performance of the reconstruction of the gamma ray energies at experimental conditions. As can be seen, the photo peak efficiency is about 3.5% at E_γ = 1000 keV and above 30% of the performance of GAMMASPHERE for energies below 2000 keV. This is the relevant energy region for the identification of characteristic gamma rays from the first excited states in the evaporation residue ^{23}Na and ^{20}Ne with $E_\gamma = 440$ keV and $E_\gamma = 1634$ keV, respectively.

3. Internal Activity Simulation

The lanthanum halides crystals are characterized by both excellent scintillation properties and intrinsic activity. The self activity originates from the radioactive isotope ^{138}La as well as contamination with the chemically equivalent isotope ^{224}Ac and its decay products. The latter component and its daughter nuclei emit alpha particles of various energies that manifest in gamma ray equivalent energies between E = 1600 keV and E = 3000 keV [12]. The self activity may seriously effect the analysis of low counting rate gamma ray measurements, while the coincidence with charged particles required in the (bigger) frame of this work is expected to completely eliminate this background contribution [5]. However, the study of the ^{138}La decay outlined in the following paragraphs serves a better characterization of the $LaBr_3$ detectors in the simulation. In addition, the pattern of the ^{138}La decay is used to progress towards an online calibration of the gamma detectors.

The half live of ^{138}La is $1.05 \cdot 10^{11}$ yr and the natural abundance is 0.0902% [13]. This introduces a background of around 90 Bq in the $LaBr_3$ crystals sized 1.5″x 2.0″ in diameter and length. The decay happens in 66.4% by electron capture (EC) into ^{138}Ba and in 33.6% by beta decay into ^{138}Ce. In both cases exclusively the first excited state (2^+) is populated leading to prompt gamma ray emission for deexcitation to the ground state (0^+) [13]. Hence, the EC and the beta decay transitions are accompanied by gamma rays with energies E_γ = 1436 keV and E_γ = 789 keV, respectively. Moreover, the Q-values in the ^{138}La decay are Q_{EC} = 1738 keV and Q_β = 1044 keV.

The end point energy of the beta decay spectrum into ^{138}Ce is E_β = 258 keV. In the present case the spectral shape is taken from a coincidence measurement described in [14]. It differs significantly from the theoretical calculation (second forbidden Gamow-Teller transition) for energies below 75 keV, but yields the best description of the shape of experimental data in the present work. As discussed in [14], this deviation may be sought with a careful evaluation of the screening factor in the theoretical calculation.

For the description of the decay into ^{138}Ba electron capture from the K and L shell are processed in the simulation with subsequent x-ray emission. The EC probabilities are taken from [15] and the energies of the edges of the K and L shell from [16]. Effects of higher order EC amount to less than 9% and are not taken into account. The $K\alpha$, $K\beta$ as well as L shell x-ray intensities and energies for barium are taken from [17] for the atomic deexcitation implementation. In summary, the gamma ray with E_γ = 1436 keV from the first excited state in ^{138}Ba and characteristic x-rays occupying the vacancies from the electron captures are generated in this decay branch. This results in a significantly broader distribution around the photo peak of the gamma ray with energy depositions up to 1473 keV where both the gamma- and the x-ray are detected. In general, two major peaks can be identified [18] and the lower contribution in commonly referred as the x-ray escape peak.

The simulation is fitted to an experimental background spectrum of a $LaBr_3$ detector sized 3.0″x 3.0″ in diameter and length. As a crude approximation a simple exponential function is chosen to account for room background and subtracted from experimental data as shown in figure 2, in the fit to the simulated ^{138}La decay scheme (blue curve). The present fit is shown as the green curve that is to be compared with the simulation in blue. The range of the fit covers all features of the ^{138}La decay and the agreement of this first approach with experimental data

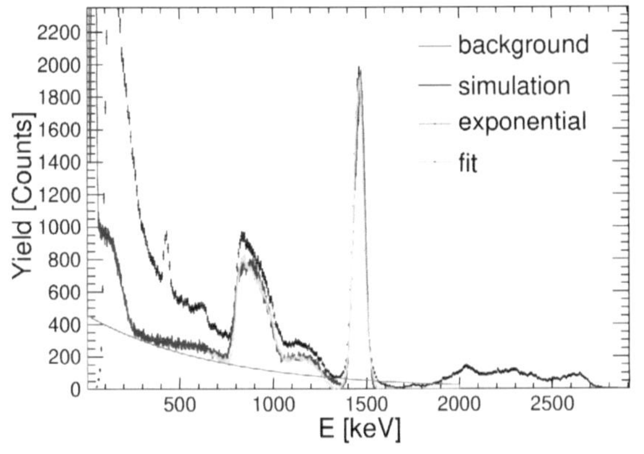

Figure 2. [color online] The fit (green curve) of the simulated decay scheme of ^{138}La (blue curve) to an experimental background spectrum (black crosses), where an exponential (red curve) is subtracted.

motivates further investigation. For energies between $E = 1600$ keV and $E = 2800$ keV a broad contribution from alpha particles emitted in the ^{224}Ac decay chain can be seen in the spectrum. The accretion of this impurities during the production of the crystals may differ from detector to detector and the description of this feature is beyond the scope of the present simulations.

The contribution at around $E = 430$ keV in figure 2 may be interpreted as the double escape peak from ^{40}K ($E_\gamma = 1461$ keV) in the borosilicate glass of the entrance window of the PM tube [18]. This would also imply a contribution around the energy region of the higher gamma line from the ^{138}La decay. Nominally, the double escape peak should appear at $E = 439$ keV and the mismatch hints on non-linear response of the $LaBr_3$ detectors towards low energy depositions [19]. Further analysis will include the characterization and correction of the energy response of the detectors as well as the simulation of the ^{40}K decay from the PM tube for the fit shown in figure 2.

For the simulation of the ^{138}La decay detailed physics input is distributed to the simulation, that reproduces the detector response to the decay features. Hence, in figure 2 also the peak positions of raw experimental data (in channels) are fitted to the simulation for instant calibration (in energy units). The performance is cross checked with ^{22}Na and ^{137}Cs runs, where the full energy depositions as well as the associated annihilation peak in the case of ^{22}Na (E_{cal}) are compared with the nominal values (E_{nom}). The result is given in table 3. The relative

Table 3. Comparison of the nominal peak positions E_{nom} with the mean energy values from the instant calibration E_{cal} utilizing characteristic lines from ^{22}Na and ^{137}Cs calibration sources. The uncertainty in percent is denoted by Δ (see text for details).

E_{nom} [keV]	E_{cal} [keV]	Δ[%]
511	526.0 ± 0.2	2.9 ± 0.2
662	672.7 ± 0.2	1.6 ± 0.2
1274	1270.5 ± 0.8	0.2 ± 0.8

uncertainty defined by $\Delta = |(E_{nom} - E_{cal})/E_{nom}|$ is in the order of a few percent. However, we aim for an absolute precision of 1 keV of the instant calibration for gamma energies between $E = 400$ keV and $E = 1800$ keV. To achieve this, the energy response of the $LaBr_3$ detectors will be characterized in ^{152}Eu runs and taken into account in the fit routine accordingly. Secondly, the origin of the classified ^{40}K contribution is investigated with dedicated simulations to obtain a reliable description of experimental data in the low energy region as well. In the end, the overall agreement of simulation and experiment in a given energy range will define the precision of the online calibration.

4. Summary

For the gamma detection in the STELLA station 36 $LaBr_3$ crystals sized 1.5″x 2.0″ will be utilized. The positioning of the detectors in the hemisphere above the target chamber is optimized utilizing Geant4 simulations. Arrangements of spherical, cylindrical, and pile stack configurations are discussed in terms of the geometrical acceptance and the photo peak efficiency. In the cylindrical setup with ten detectors in the lowest line the geometrical acceptance is 23.1% and the photo peak efficiency at $E = 1$ MeV is around 3.5%. The results are backed by the calculation of pair-wise distance distributions reflecting the characteristics of the detectors arrangements from the simulations.

Based on the activity of ^{138}La the performance of the current simulation is tested qualitatively. This comprises the description of the interactions of electrons and gamma rays with the scintillating material as well as the detector geometry implementation. The experimental background spectrum of a 3″x 3″ $LaBr_3$ crystal, that is dominated by the ^{138}La decay, is reproduced in the energy range between $E = 650$ keV and $E = 1650$ keV with a general exponential to account for exterior background sources. The fit is also used for an instant calibration of the gamma detectors for the STELLA-FATIMA project and the absolute uncertainty of the online energy calibration routine is 15 keV in this first approach.

Acknowledgments

This work was supported by the IdEx Attractivity grant (F) of the University of Strasbourg, by the STFC (UK) grant number ST/L005743/1, and by the DFG (GER) through grant SFB 1245.

References

[1] Jenkins D G and Courtin S 2015, *Phys. Jour.* G **42**, 034010
[2] Chieffi A, Limongi M and Straniero O 1998, *ApJ*, **502**, 737
[3] Gasques L R, Brown E F, Chieffi A, Jiang C L, Limongi M, Rolfs C, Wiescher M and Yakovlev D G 2007, *PRC* **76**, 3, 035802
[4] Courtin S *et al.* 2016, NIC2016, accepted *JPS Conf. Proc.*
[5] Jiang C L *et al.* 2012, *NIM* A **682**, 12
[6] Patterson J R, Winkler H and Zaidins C S 1969, *APJ* **157**, 367
[7] Erb K A, Betts R R, Korotky S K, Hindi M M, Tung P P, Sachs M W, Willett S J and Bromley D A 1980, *PRC* **22**, 22
[8] Roberts O J, Bruce A M, Regan P H, Podolyák Z, Townsley C M, Smith J F, Mulholland K F and Smith A 2014, *NIM* A **748**, 91
[9] Lorusso G *et al.* 2016, *Appl. Rad. Iso.* **109**, 507
[10] Lauritsen T 2016, <http://www.phy.anl.gov>
[11] Lee I Y 1990, *Nucl. Phys.* A **520**, 641c
[12] Milbrath B D, Runkle R C, Hossbach T W, Kaye W R, Lepel E A, McDonald B S and Smith L E 2005, *NIM* A **547**, 504
[13] Firestone R B and Shirley V S 1997, *Table of Isotopes*, (Wiley)
[14] Quarati F G A, Khodyuk I V, van Eijk C W E, Quarati P and Dorenbos P 2012, *NIM* A **683**, 46
[15] Bé M M *et al.* 2010, *Monographie BIPM* **5**, Table of Radionuclides, (BIPM)
[16] Deslattes R D, Kessler Jr. E G, Indelicato P, de Billy L, Lindroth E and Anton J 2003, *Rev. Mod. Phys.* **75**, 35
[17] Chu S Y F, Ekström L P and Firestone R B 2016, *The Lund/LBNL Nuclear Data Search*, <http://nucleardata.nuclear.lu.se/toi/index.asp>
[18] Camp A, Vargas A and Fernández-Varea J M 2016, *Appl. Rad. Iso.* **109**, 512
[19] Giaz A *et al.* 2013, *arXiv* 1308.6085

Measurement of fast neutron detection efficiency with ^{6}Li and ^{7}Li enriched CLYC scintillators

A Mentana[1,2], F Camera[1,2], A Giaz[1], N Blasi[1], S Brambilla[1], S Ceruti[1,2], L Gini[1], F Groppi[1,2], S Manenti[1,2], B Million[1] and S Riboldi[1]

[1] INFN Milano, Via Celoria, 16, 20133 Milano, Italy
[2] University of Milano, Via Celoria, 16, 20133 Milano, Italy

E-mail: alice.mentana@mi.infn.it

Abstract. The CLYC (Cs_2LiYC_6:Ce) crystal belongs to the elpasolite scintillator family, discovered about 15 years ago. It is a very interesting material because of its good energy resolution and its capability to identify and measure gamma rays and fast/thermal neutrons. In the present work, the fast neutron detection efficiency for two different CLYC cylindrical samples has been measured. These two crystals, both with dimension (thickness x diameter) 1"x1", were respectively enriched with more than 99% of ^{7}Li (CLYC-7) and with $\sim$ 95% of ^{6}Li (CLYC-6). The presence of the ^{6}Li isotope makes the CLYC-6 ideal to detect thermal neutrons. In order to compare the two scintillators, only the detection efficiency for fast neutrons was considered, neglecting the energy region associated to thermal neutrons in both the crystals. The measurement was performed at the L.A.S.A. Laboratory of INFN and University of Milano (Italy), using a ^{241}Am-Be source.

1. Introduction

The elpasolite crystals were discovered approximately 15 years ago, and are excellent in terms of gamma and neutron detection [1]-[4]. Crystals identified as CLLB:Ce ($Cs_2LiLaBr_6$:Ce), CLLC:Ce (Cs2LiLaCl6:Ce) and CLYC:Ce belong to this new class of detectors. All these materials are suitable to identify and measure the energy of gamma rays and neutrons. The very different scintillation light decay response to different types of radiation allows them to clearly identify the incident radiation using the Pulse Shape Discrimination (PSD). The signal produced by the gamma rays contains the fast component of the scintillation light (decay time $\sim$ few ns) and the slowly decaying components channelled through the Ce^{3+} ions. The neutron signal, instead, does not have a fast component and it is characterized by a slower decay time constant [5]. Furthermore, the CLYC scintillators can detect both thermal and fast neutrons. Fast neutrons are detected using the reaction $^{35}Cl(n,p)^{35}S$ and $^{35}Cl(n,\alpha)^{32}P$ [4] where the proton or alpha particle energy is linearly related to the energy of the neutron. The thermal neutrons detection capability arises from the presence of ^{6}Li, with the reaction $^{6}Li(n,\alpha)t$ which has a cross-section of 940 barns.

2. The measurement

Aim of the work was to estimate the fast neutron detection efficiency of two CLYC crystals, respectively enriched with more than 99% of ^{7}Li (CLYC-7) and around 95% of ^{6}Li (CLYC-6).

The measurement was performed at L.A.S.A. Laboratory of I.N.F.N. and University of Milano in 2015, using a calibrated ^{241}Am-Be source, which generates neutrons via the following reaction:

$$^{241}Am \rightarrow {}^{237}Np + \alpha + \gamma, \tag{1}$$

$$^{9}Be + \alpha \rightarrow {}^{13}C^{*} \rightarrow {}^{12}C^{*} + n. \tag{2}$$

The energy spectrum of the emitted neutrons is a continuum up to around 10 MeV (see Fig. 1, [6]).

To calculate the fast neutron detection efficiency $\varepsilon_{n_{fast}}$, defined as the ratio between the fast neutron flux measured $\phi_{n_{fast}}$ and the one expected $\phi_{expected}$:

$$\varepsilon_{n_{fast}} = \frac{\phi_{n_{fast}}}{\phi_{expected}}, \tag{3}$$

one needs to count the number of fast neutrons per second interacting in the crystal ($\phi_{n_{fast}}$). The expected flux is obtained knowing the source activity A_n and the solid angles the crystal subtends, Ω ($\phi_{expected} = A_n \cdot \Omega$).
An accurate interpretation of the neutron experimental spectrum, which allows the selection of fast neutron events, is based on the detailed understanding of the neutron interaction in a CLYC scintillator.

2.1. n interaction in CLYC crystals

In CLYC-6 and CLYC-7, the fast neutron interacts with ^{35}Cl via two reactions: $^{35}Cl(n,p)^{35}S$ and $^{35}Cl(n,\alpha)^{32}P$. In CLYC-6, the ^{6}Li presence allows the detection of thermal neutron also, through $^{6}Li(n,\alpha)t$. These three reactions are characterized by three different Q-values and the emitted particles have different quenching factors f_q. Realizing that the energy measured by the crystal E_{eff}, in terms of MeVee (i.e. the scintillation light produced), is expected to be:

$$E_{eff} = (E_n + Q_{val}) \cdot f_q, \tag{4}$$

a fixed E$_n$ value (the real neutron energy, in MeV) will correspond to three different E_{eff} values, depending on the reaction the neutron undergoes. In other words, the total experimental spectrum of neutrons from the ^{241}Am-Be source is obtained from the sum of three (or two for CLYC-7) different spectra, each one associated to a reaction, with a characteristic Q_{val} and f_q. These spectra are shown respectively in Fig. 3 for $^{35}Cl(n,p)^{35}S$, in Fig. 4 for $^{35}Cl(n,\alpha)^{32}P$ and in Fig. 2 for $^{6}Li(n,\alpha)t$ reaction. Actually, each component needs also to be weighted on its reaction cross section. The cross section curves relative to the three neutron absorption reactions are reported in Fig. 5.
To summarise, each neutron induced reaction produces a spectrum as the one of Fig. 1 shifted by the Q_{val}, scaled by the quenching factor and weighted on the cross section curves (see Fig. 6).

2.2. Fast neutron efficiency estimation

Dealing with a CLYC-7, the total ^{241}Am-Be neutron spectrum arises from the sum of the blue and red spectra of Fig. 6 (due to n interaction in ^{35}Cl). For a CLYC-6 crystal, also the green spectrum contribution of Fig. 6 has to be included (due to interaction in ^{6}Li).
Looking at Fig. 5, it is clear how the green cross section curve associated to neutron capture in ^{6}Li dominates (it is about 1000 times higher than the other curves) in the region 3.1 MeVee $< E_{eff} <$ 3.6 MeVee, which corresponds, for these reaction, to the thermal neutron energy region. In order to be able to compare fast neutron detection efficiency for the two crystal, this region will be neglected for both.

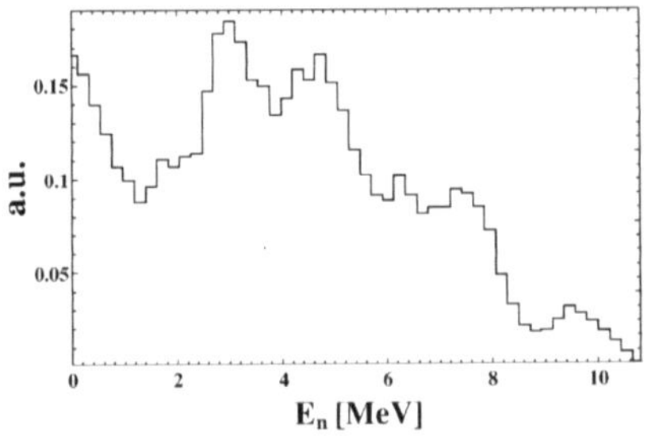

Figure 1. Neutron spectrum of a ^{241}Am-Be source [6].

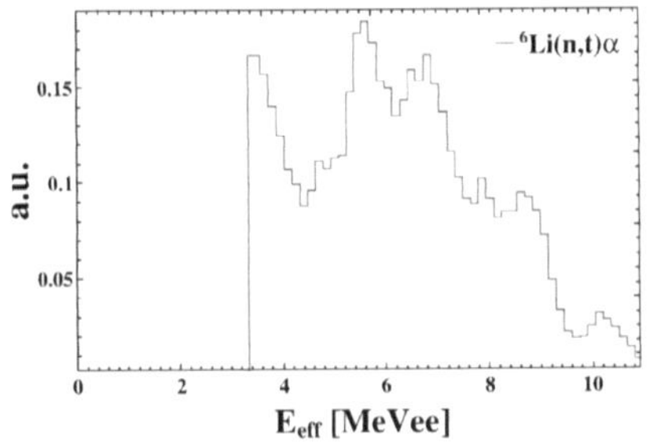

Figure 2. Neutron spectrum of a ^{241}Am-Be source as a function of the scintillation light produced, i.e. the effective energy measured E_{eff} (in MeVee), due to the reaction $^{6}Li(n,\alpha)t$ (only in CLYC-6).

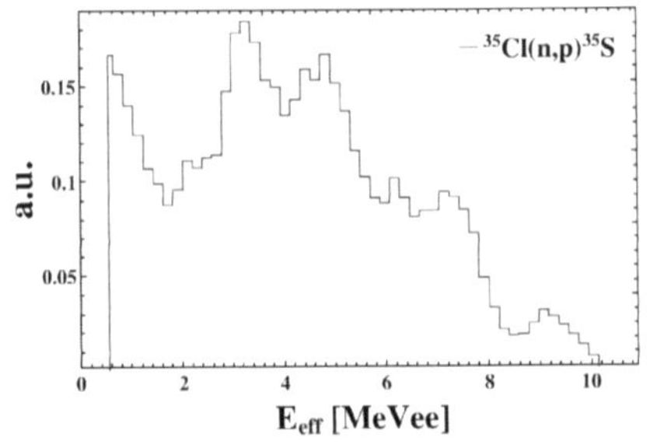

Figure 3. Neutron spectrum of a ^{241}Am-Be source as a function of E_{eff}, due to the reaction $^{35}Cl(n,p)^{35}S$ (CLYC-6 and CLYC-7).

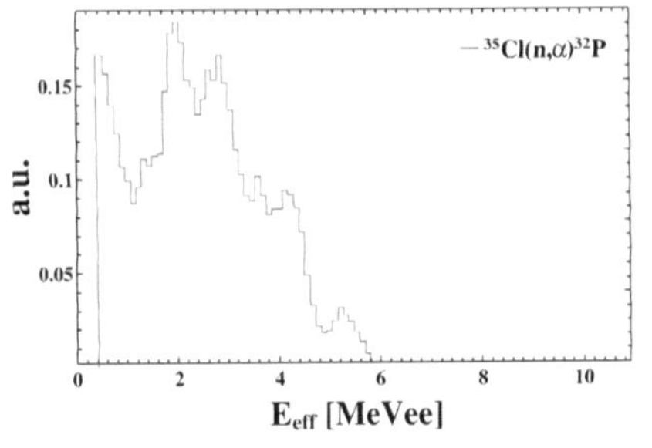

Figure 4. Neutron spectrum of a ^{241}Am-Be source as a function of E_{eff}, due to the reaction $^{35}Cl(n,\alpha)^{32}P$ (CLYC-6 and CLYC-7).

Counting the number of neutrons in the two energy regions, 0-3.1 MeVee and 3.6-10 MeVee, the measured fast neutron flux $\phi_{n_{fast}}$ is obtained. The preliminary efficiency values, for CLYC-6 and CLYC-7, derived (eq. 3) are listed in Tab. 1.

A direct comparison between the efficiency values obtained for the two different crystals is not possible over all the energy range, due to the presence, in CLYC-6 only, of ^{6}Li. Only below 3.1 MeVee, indeed, the ^{6}Li contribution is not present. Such a contribution to the efficiency, in the

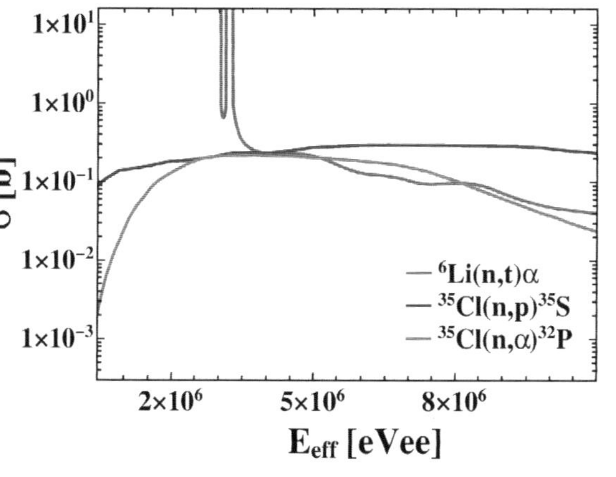

Figure 5. Comparison between the cross section curves of the three different neutron absorption reaction in terms of E_{eff}. The green curve refers to $^{6}Li(n,\alpha)t$ (in CLYC-6 only), the red one refers to $^{35}Cl(n,\alpha)^{32}P$ and the blue one, that dominates below 1 MeVee, refers to $^{35}Cl(n,p)^{35}S$.

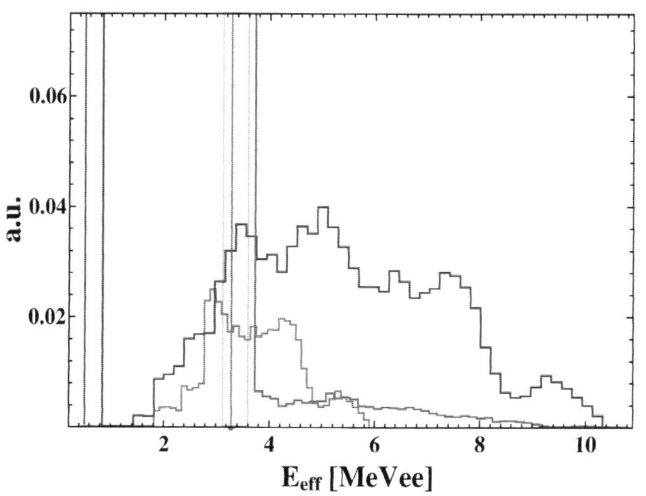

Figure 6. ^{241}Am-Be neutron predicted spectra as a function of E_{eff} weighted on the cross section for the three different neutron absorption. The blue and red spectra are referred to n interaction with ^{35}Cl, respectively $^{35}Cl(n,p)^{35}S$ and $^{35}Cl(n,\alpha)^{32}P$. The green spectrum is associated to $^{6}Li(n,\alpha)t$. The region between 3.1 MeVee and 3.6 MeVee (yellow rectangle) is completely dominated by the thermal neutrons peak (due to the interaction in ^{6}Li).

region over 3.6 MeVee, can be estimated taking into account the amount of Li in a Cs_2LiYC_6:Ce crystal, the isotopic abundance of ^{6}Li, the AmBe source intensity and the cross section. In this case, the ^{6}Li contribution to efficiency over 3.6 MeV is around 15%.

Table 1. The intrinsic values of fast neutron detection efficiency measured for CLYC-6 and CLYC-7.

	$\varepsilon_{n_{fast}}(\%)$
CLYC-6	(1.1 ± 0.1)
CLIC-7	(0.7 ± 0.1)

3. Conclusion

The values of intrinsic fast neutron detection efficiency has been estimated for two CLYC crystal samples, one enriched with ^{6}Li (~95%) and one with ^{7}Li (more than 99%). Such values are referred to the whole energy range, neglecting only the region associated to the interaction of thermal neutrons with ^{6}Li.

The preliminary results of this work are consistent with the ones recently published in [7] (simulation) and [8] (measurement), even though these are relative to efficiency as a function of the energy to CLYC crystal with a different relative abundance of ^{6}Li and ^{7}Li.

The forthcoming step will be to measure the fast neutron detection efficiency for the same CLYC crystals and for a larger one (2"x2" in size (thickness x diameter)) as a function of the neutron energy. For this purpose, a significant improvement in the fast neutron identification will be provided by the combined use of energy signal (the PSD technique, as in the present work) and time of flight measurement [9].

4. References

[1] Combes C M *et al* 1999 *Journal of Luminescence* B **82** pp 299-305
[2] Glodo J *et al* 2008 *IEEE Transactions on Nuclear Science* **55**,3 p 1206
[3] Glodo J *et al* 2011 *IEEE Transactions on Nuclear Science* **58**,1 p 333
[4] Smith B *et al* 2013 *IEEE Transactions on Nuclear Science* **60**,2 p 855
[5] `http://rmdinc.com/about-us/dynasil-corporate-profile/`
[6] Usha Pujala *et al* 2011 *Radiation Protection and Envirroment* **34** pp 262-266
[7] D'Olympia N *et al* 2014 *Nuclear Instrumental and Methods in Physisce Research A* **763** pp 433-441
[8] Smith B *et al* 2015 *Nuclear Instrumental and Methods in Physisce Research A* **784** pp 162-167
[9] Giaz A *et al* 2016 *Nuclear Instrumental and Methods in Physisce Research A* **825** p 51

Neutron-gamma discrimination via PSD plastic scintillator and SiPMs

M P Taggart[1], C Payne[1], and P J Sellin[1]
[1] Department of Physics, University of Surrey, Guildford GU2 7XH, UK
E-mail: m.taggart@surrey.ac.uk

Abstract. The reduction in availability and inevitable increase in cost of traditional neutron detectors based on the ^{3}He neutron capture reaction has resulted in a concerted effort to seek out new techniques and detection media to meet the needs of national nuclear security. Traditionally, the alternative has been provided through pulse shape discrimination (PSD) using liquid scintillators. However, these are not without their own inherent issues, primarily concerning user safety and ongoing maintenance. A potential system devised to separate neutron and gamma ray pulses utilising the PSD technique takes advantage of recent improvements in silicon photomultiplier (SiPM) technology and the development of plastic scintillators exhibiting the PSD phenomena. In this paper we present the current iteration of this ongoing work having achieved a Figure of Merit (FoM) of 1.39 at 1.5 MeVee.

1. Introduction

It is well-known that the emission from organic scintillation detectors comprises both fast and slow components. For the vast majority of applications it is the prompt emission that is most useful. However, pulse shape discrimination is a special case in that it is the slow component of emission which is of greatest importance. For scintillators exhibiting the PSD phenomena the delayed fluorescence can be directly related to the type of incident radiation[1], i.e. it can be used to distinguish between pulses caused by neutrons and gamma rays in a way that other scintillators cannot. Thus a PSD system can help end the reliance on a dwindling supply of ^{3}He for national nuclear security.

PSD is most strongly observed within the category of organic liquid scintillators such as EJ-309[2, 3, 4]. The chemical properties of these materials make them undesirable for mass deployment, especially in situations where they would require regular maintenance by personnel. Conversely, plastic scintillators are known for their stability and ruggedness, making them a far more suitable prospect for a long-term deployment with possible public exposure. With the development of EJ-299-33 [5], a plastic scintillator exhibiting PSD properties, there appears to be a solution to the absence of neutron detection capability.

All scintillators require a coupled sensor to detect the emission of light, a role generally filled by the photomultiplier tube (PMT). Unfortunately, PMTs usually require a high voltage power supply, presenting an electrical hazard to their deployment, as well as the nature of their operation being such that they remain unsuitable for use with magnetic fields[6]. The silicon photomultiplier (SiPM) is an alternative which solves these issues[7], and with recent technological improvements the high dark current of previous generations has been greatly reduced[8].

Journal of Physics: Conference Series **763** (2016) 012007 doi:10.1088/1742-6596/763/1/012007

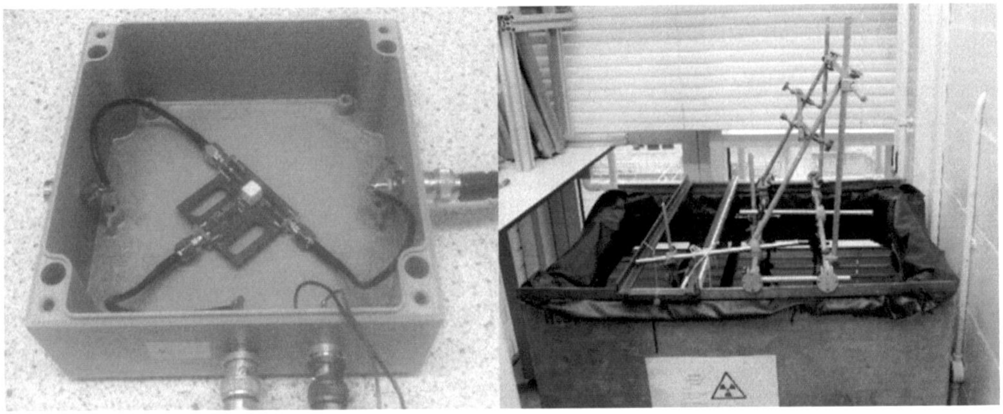

Figure 1. The EJ-299/SiPM detector system (left), the AmBe neutron source tank (right).

The use of digital electronics via fast waveform digitisers enables the raw SiPM pulse to be processed and subsequently analysed. Ongoing development work in the field of digital electronics has allowed a study on the relative importance of various pulse processing factors, specifically Analogue-to-Digital Conversion (ADC) resolution[9]. This paper concerns the ongoing development of a digitised plastic scintillator-SiPM PSD system and we present new Figures of Merit (FoM) in comparison to those established with our previous generation data acquisition (DAQ)[10].

2. Experimental Summary

The detector combined a small (6x6x6 mm^3) piece of EJ-299 plastic scintillator and a SensL MicroFC-SMA-60035 C-series SiPM. To maximise light transmission the scintillator had been highly polished, painted with EJ-510 titanium-oxide paint, and optically coupled to the SiPM with standard silicon grease. In addition, the scintillator's wavelength of peak emission corresponds well to the peak response of the SiPM, which are both at 420 nm[8, 11].

The detector setup consisted of a sealed, light-tight metal box with surface mount BNC ports for access to the fast/slow outputs of the SiPM, the bias supply, and an additional connector to monitor the temperature via a PT100 platinum thermistor. The detector housing is shown in Figure 1. The SiPM selected was the largest single-cell device currently manufactured by SensL at 6x6 mm^2 with 18980 microcells at a 64% fill factor, which enables the detector to be sensitive over a greater dynamic range than smaller devices. The breakdown voltage (Vbr) was found to be in accordance with datasheet values at 24.8 V (at 26°C) using a Keithley 487 picoammeter driven by LabView operating software. The operating voltage was selected as Vbr + 2.5 V and was provided by an Ortec 710 Quad Bias Supply.

Although access to both outputs of the SiPM was available, only the slow output was used, which was connected to a Caen digitiser. Two digitisers were used, a 10-bit V1751 (bandwidth 500 MHz and sample rate of 1 GS/s), and a 14-bit V1730C (bandwidth 250 MHz and sample rate 500 MS/s); previous work at the University of Surrey has established ADC resolution to be one of the most important factors for achieving good quality PSD with digital systems[9]. Pulses were processed either through the on-board Digital Pulse Processing (DPP) PSD algorithms, allowing the digitisers Field Programmable Gate Array (FPGA) to perform the charge integration, or the waveform was simply recorded for post-processing with external, in-house analysis software.

The mixed radiation field produced by a 17.57 GBq AmBe source was used to test the PSD capability. The use of fast neutrons was enabled through insertion of an air-tube into the 1 m^3 water tank shielding the AmBe source. The detector was placed 50 cm from the source, but

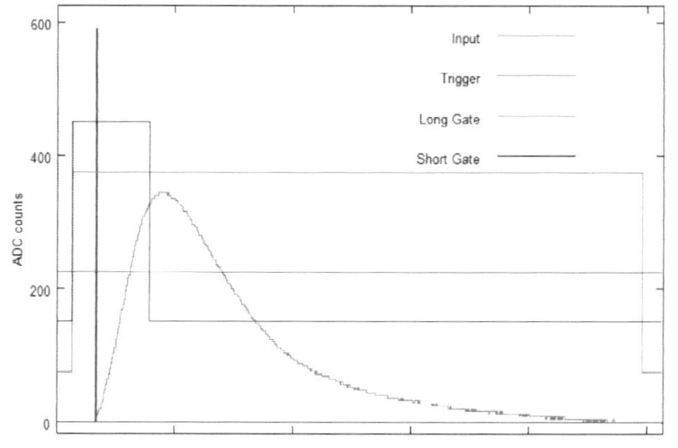

Figure 2. A typical radiation waveform (red), with the trigger, short gate, and long gate overlaid.

less than 1 cm of water shielding remained. The AmBe source has a significant gamma-ray flux, therefore 5 cm of lead was placed between the source and detector to avoid over-saturation due to gamma signals. The setup can be seen in Figure 1.

2.1. Charge integration

The charge integration technique for PSD involves the application of two time gates to each pulse and establishing the ratio between them. As it is the slow component that contains the information regarding the nature of the incident radiation[1] it is the relative strength of this that allows neutron-gamma discrimination to take place.

The application of the two time gates is depicted in Figure 2, with the charge collected from each integration used to determine the PSD parameter as given by:

$$PSD = 1 - \frac{Q_s}{Q_l} \tag{1}$$

where Q_s and Q_l are the charge integrated in the short and long gates respectively. The optimum short gate was found to be 66 ns with the V1751 while the long gate encompassed the entire record length (700 ns) of the pulse. For direct comparison, these settings were maintained for the V1730C measurements.

3. Results

The standard method of determining the quality of neutron-gamma discrimination is via a 2-D histogram of total integrated charge versus the respective PSD parameter of that pulse. The PSD parameter is calculated by equation 1, and the integrated charge is simply the Q_l value of that equation. This histogram for the C-series/V1730C system is shown in Figure 3, and exhibits the characteristic PSD shape of two, well-defined regions, where the upper region contains neutron events and the gamma pulses are confined to the lower region.

The quantitative performance of PSD for a given detector system is defined by a Figure of Merit (FoM). The FoM details the neutron-gamma discrimination for a given energy threshold, and is a function of both the separation and resolution of peaks in the PSD parameter, i.e. a projection of the y-axis of Figure 3. The widely used FoM calculation[12] here is in common with other recent works [10, 13], and is given by:

$$FoM = \frac{S}{\Gamma_1 + \Gamma_2} \tag{2}$$

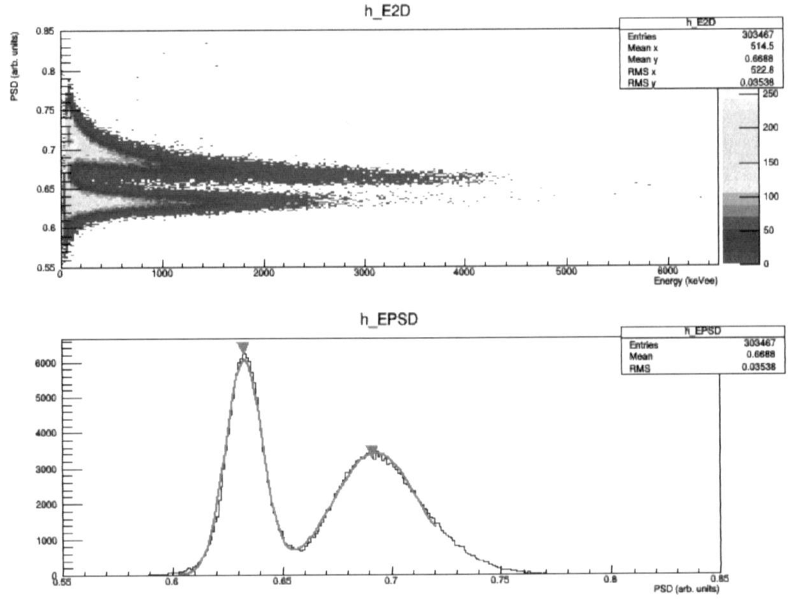

Figure 3. The standard PSD histogram for the C-series SiPM with V1730C digitiser. The PSD projection has had a double Gaussian function fitted to enable extraction of the parameters for the calculation of the Figure of Merit.

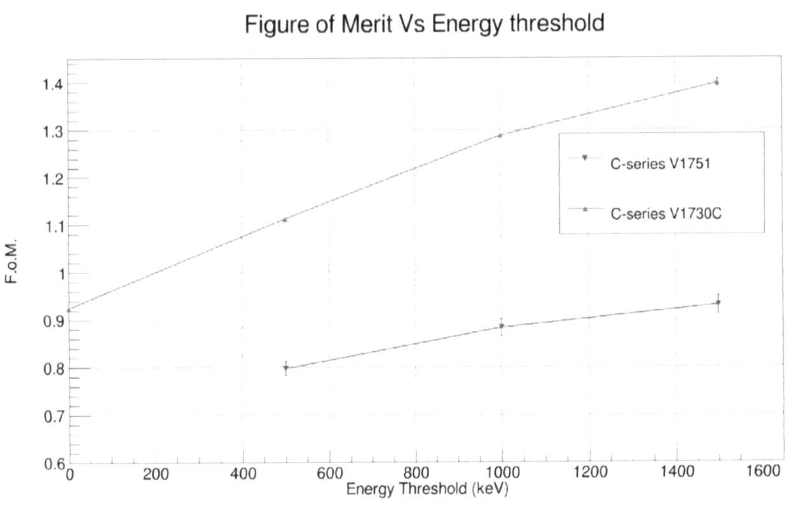

Figure 4. Figures of Merit versus energy for C-series SiPM.

where S is the separation between the neutron and gamma peaks and Γ_1 and Γ_2 are their full widths at half maximum.

There is generally a strong correlation between the FoM and increasing neutron/photon energy. At the lowest energies the FoM is significantly reduced due to overlap between neutron and gamma ray PSD distributions. The FoM as a function of energy threshold is shown in Figure 4.

4. Conclusion

The presented system has achieved significant separation of the neutron and gamma ray pulses from an AmBe source, achieving a FoM of 1.39 at 1.5 MeVee. In particular we have shown the substantial improvement to be made when increasing the resolution of the signal pulse height as determined by the digitiser specification. The variation in neutron and gamma pulses with EJ-299-33 is subtle, and the amplitudes are small. The V1730C 14-bit digitiser has a resolution of ~0.03 mV/ADC unit. A typical difference between neutron and gamma pulses obtained by the V1730C would be ~100 ADC units over each sample of a pulse, corresponding to ~3 mV. Similar pulses processed by a V1751 10-bit digitiser would be separating based on a difference of ~3 ADC units per sample. We believe that it is this fundamental difference that produces superior PSD for the V1730C over the V1751. Furthermore, the advanced technologies implemented at this stage have shown that a plastic scintillator/SiPM system can compete in terms of performance with alternative PSD systems. The work of Liao and Yang[13] highlighted the potential of an EJ-299-33/SiPM system, quoting FoMs of 0.59 and 0.76 at 1 MeVee for a charge integration and complex algorithm respectively. The system in [13] however, used an older B-series device. Contrastingly, Cester *et al.*[14] confirms the superior PSD qualities of PMT-based systems and makes use of a V1751 10-bit digitiser achieving a FoM of 1.3 at 1 MeVee. For comparison to other scintillation materials, Yanagida *et al.* quote a FoM of 1.53 for an anthracene/PMT system[15], with Woolf *et al.* suggesting that a liquid scintillator system has a FoM of around 2[16].

At present the SiPM technology is undergoing continued development and new versions of the sensor are available. It is likely that the J-series device will see further improvements in its signal-to-noise ratio and that the FoM will increase as a result. The next stage of work will be to repeat the investigation presented here with the J-series SiPM, and then to investigate the tolerance of PSD performance with respect to the sample rate of the digitiser by down-sampling the waveform. It has already been shown that the increased resolution of the V1730C (which allows for the subtleties between the neutron and gamma pulses to be more readily discerned) more than compensates for the reduction in sample rate from the V1751, but the limitations of this trade-off are yet to be determined.

References

[1] Knoll G F 2000 *Radiation Detection and Measurement* (Wiley)
[2] Pozzi S A, Clarke S D, Flaska M and Peerani P 2009 *Nucl. Instr. Meth. A* **608** 310–315
[3] Clarke S D, Flaska M, Pozzi S A and Peerani P 2009 *Nucl. Instr. Meth. A* **604** 618–623
[4] Flaska M and Pozzi S A 2007 *Nucl. Instr. Meth. A* **577** 654–663
[5] Zaitseva N, Rupert B L, Pawelczak I, Glenn A, Martinez H P, Carman L, Faust M, Cherepy N and Payne S 2012 *Nucl. Instr. Meth. A* **668** 88–93
[6] Christensen N L, Hammer B E, Heil B G and Fetterly K 1995 *Phys. Med. Biol.* **40** 691–697
[7] Hawkes R, Lucas A, Stevick J, Llosa G, Marcatili S, Piemonte C, Guerra A D and Carpenter T A 2007 *IEEE NSS MIC Conf. Rec.* vol M18-118 pp 3400–3403
[8] C-series family `www.sensl.com/products/c-series/` accessed: 20016-06-03
[9] Payne C 2015 Ph.D. thesis University of Surrey
[10] Taggart M P, Ellis M, Payne C and Sellin P J 2015 *Proc. Defence and Security Doctoral Symp.*
[11] Ej-299-33 `www.southernscientific.co.uk/catalog/products/EJ-299-34` accessed: 20016-06-03
[12] Winyard R A, Lutkin J E and McBeth G W 1971 *Nucl. Instr. Meth.* **95** 410–153
[13] Liao C and Yang H 2015 *Nucl. Instr. Meth. A* **789** 150–157
[14] Cester D, Nebbia G, Stevanato L, Pino F and Vesti G 2014 *Nucl. Instr. Meth. A* **735** 202–206
[15] Yanagida T, Watanabe K and Fujimoto Y 2015 *Nucl. Instr. Meth. A* **784** 111–114
[16] Woolf R S, Hutcheson A L, Gwon C, Phlips B F and Wulf E A 2015 *Nucl. Instr. Meth. A* **784** 80–87

Journal of Physics: Conference Series **763** (2016) 012008 doi:10.1088/1742-6596/763/1/012008

Applications for New Scintillator Technologies in Gamma Ray Astronomy

Mark L. McConnell[1,2], Peter F. Bloser[1], Jason Legere[1], James M. Ryan[1]

[1]Space Science Center, University of New Hampshire, Durham, NH 03824 USA

[2]Dept. of Earth, Oceans, and Space, Southwest Research Institute, Durham, NH 03824 USA

E-mail: mark.mcconnell@unh.edu

Abstract. Scintillators have long been used for probing the high energy universe. The reliability and low cost of scintillator-PMT detectors have made them the de facto standard for experiments on high altitude balloons and in orbiting satellites. New scintillators and new readout technologies offer important opportunities for more capable experiments. Recent scintillator developments include faster signals, increased light output, improved energy resolution, and better handling characteristics. Although PMTs remain effective for scintillator readout, new technologies offer more compact, rugged devices with much lower operational voltages. The adoption of these technologies is not without its difficulties, especially for space applications, where the technology readiness level can be an important consideration. To illustrate these issues, we will discuss the use of scintillators in Compton imaging experiments. At energies from about 500 keV to 30 MeV, Compton telescopes are the most effective means of imaging the gamma ray sky. To date, the only Compton telescope that has flown in space was the COMPTEL instrument on NASA's Compton Gamma Ray Observatory (CGRO). CGRO, launched in 1991 and de-orbited in 2000, was based entirely on the use of technologies from the 1980's. We have been working on an improved Compton telescope design, called the Advanced Scintillator Compton Telescope (ASCOT). It is much like COMPTEL, but utilizes up-to-date scintillator and readout technologies.

1. Introduction

Gamma-ray astronomy is not easy. First, it is necessary to observe from above the atmosphere. Second, the very low flux of photons from celestial γ-ray sources means that the combination of observation time and effective area must be large. Getting large instruments into space is a challenge. Third, orbiting instruments operate in a very high radiation environment. The radiation consists of cosmic rays plus the by-products of cosmic ray interactions with the Earth and with the spacecraft itself. The detector itself can even become a source of radiation. Consequently, the instrumental background can be quite high. Scintillator technology has long played a crucial role in γ-ray astronomy. Scintillators are a proven technology that can often be fabricated in large areas at relatively low cost. Although their energy resolution may be somewhat limited, there are many advantages to scintillation detectors (in particular, their timing characteristics) that often make them the detector of choice for many applications in astrophysics. Here we review some of the ways in which scintillators have been used in astrophysics and then discuss some of the work we have done utilizing some of the scintillator technologies.

2. Gamma Ray Telescopes

Within the energy range from above 10 keV up to several hundred GeV, the range covered by orbiting γ-ray telescopes, there are three primary modes of photon interaction – photoelectric absorption, Compton scattering and pair production. For imaging of γ-rays, there are three basic designs, each of which is tailored to a particular interaction process.

2.1. Coded Mask Imaging

At energies below $\sim$ 1 MeV, the regime where the photoelectric effect is the dominant photon interaction mechanism, the typical approach to γ-ray imaging is the use of a coded mask [1]. Imaging is achieved by creating a shadow of an opaque mask (opaque to the photons of interest), which must then be measured using some position-sensitive detection plane. An accurate measure of the projected shadow allows for the reconstruction of the incident flux distribution. The requirement of having an opaque mask with reasonable geometry (not too thick so as to distort off-axis imaging) limits the energy range to below 1 MeV. In addition, the presence of a passive mask in a high radiation environment can also increase the instrument background due to activation effects. Nonetheless, coded mask telescopes have proven to be especially effective at energies below a few hundred keV. Scintillators have often been used for the position-sensitive detection plane. Discrete detector elements have sometimes been used, but location sensitive scintillation detectors have also been employed. Notable examples of orbiting coded mask instruments include SIGMA/GRANAT [2], INTEGRAL/IBIS [3], INTEGRAL/SPI [4], and Swift/BAT [5].

2.2. Compton Imaging

In the regime where Compton scattering is the dominant interaction process (roughly between 1 and 30 MeV), Compton imaging is employed. This technique takes full advantage of the Compton scattering process by measuring the kinematics of each Compton scattered photon. The measurement of the Compton kinematics for a single photon event requires measuring the location and energy loss at the site of the Compton scatter. It also requires a measure of the scattered photon energy and direction, which can be ascertained by a second detector designed to absorb the scattered photon. The arrival direction of the incident photon can be determined to within some annulus on the sky. The angular radius of this "event circle" is determined by the measured Compton scatter angle and the width of the event circle is determined, in part, by the energy resolution. An analysis of overlapping event circles can be used to determine the photon arrival direction. The only dedicated Compton telescope flown to date was the Compton Imaging Telescope (COMPTEL) on the Compton Gamma-Ray Observatory (CGRO) [6].

2.3. Pair Production Imaging

Pair production telescopes rely on the conversion of an incident photon into an electron-positron pair. Accurate tracking of pairs is required to determine the arrival direction of the incident photon (based on the initial momentum vectors of the outgoing electron and positron) and the incident photon energy (from the opening angle between the electron and positron momentum vectors). Scattering of the electron and positron distorts the tracks and limits the performance. This can be an especially significant problem at lower energies. Typically, track chambers or solid state strip detectors are used for track measurement. Scintillation detectors, however, can play an important role. Large volume scintillators (with limited spatial resolution) are often used as calorimeters to insure a full energy measurement of the electron and positron. Notable examples of orbiting pair telescopes include SAS-II, COS-B [7], CGRO/EGRET [8], Fermi/LAT [9], and AGILE [10].

3. MeV Astronomy

Historically, the field of γ-ray astronomy has grown over the past 50 years with a series of balloon and spacecraft instruments. These efforts have led to a series of successful orbital missions that have included the Compton Gamma-Ray Observatory (CGRO) [11], INTEGRAL [12], and Fermi. The scientific results from these missions have revolutionized our understanding of the γ-ray sky. Collectively, however, these instruments have not provided a uniform sensitivity coverage of the γ-ray spectrum. In particular, the region from about 1 MeV up to about 50 MeV has been poorly studied relative to neighboring energy regimes. The reduced sensitivity coverage in this part of the spectrum is sometimes referred to as the "MeV Gap." As astronomers are now looking to plan future γ-ray missions, the need to fill in this gap with more sensitive instrumentation is widely recognized as one of the next strategic goals. This particular part of the spectrum covers the domain of the Compton telescope. It also encompasses the lower energy part of the pair production regime, where scattering of the electron-positron pair limits the angular resolution, and therefore the sensitivity, of pair production imaging. It also corresponds to a part of the spectrum that is experimentally challenged by significant levels of background. Most γ-ray lines from nuclear de-excitation can be found in the energy range of roughly 1–10 MeV, the so-called "nuclear line region." On orbit, the intense charged particle background results in the activation of many materials. Consequently, the spacecraft and the instrument itself literally glow in γ-rays. A suitable choice of materials (e.g., carbon fiber material rather than aluminum) can help limit the induced background, but the instrument itself must still be capable of recognizing and rejecting these background sources.

Over the years, several efforts have been made to develop a next-generation Compton telescope. Some have been based on solid state detector technologies, such as Ge or Si-strip detectors [13, 14, 15]. Others have explored the use of time-projection-chambers [16, 17]. Our work leverages our past experience with the COMPTEL experiment and has focused on the use of scintillator-based designs, recognizing the important capabilities of new types of scintillator material couple with new readout technologies.

3.1. CGRO/COMPTEL

The COMPTEL instrument was designed to image γ-rays in the 1–30 MeV energy range [6]. The principle of operation is illustrated in Fig. 1. It consisted of two layers of scintillation detectors. The upper (D1) detector layer (Fig. 2) consisted of seven large volume liquid scintillator (NE 213A) tanks (28.5 cm in diameter by 8.5 cm thick), each read out by an array of eight PMTs. The lower (D2) detector layer (Fig. 3) consisted of fourteen large volume NaI(Tl) scintillator crystals (28.2 cm in diameter and 7.5 cm thick) read out by an array of seven PMTs. In each case, signals from the PMT array were used to localize the event interactions to within 1-2 cm. The two detector layers (each surrounded by plastic anti-coincidence shielding to reject charged particles) were separated by distance of 1.5 m. Incoming photons would Compton scatter in the low-Z liquid scintillator in the D1 layer. The interaction location and energy deposit would be measured in the D1 detector. The scattered photon would subsequently travel downward and be fully absorbed by an interaction in D2. Once again, the location of the interaction and the total energy loss were measured.

The COMPTEL design incorporated two key features that were designed to help identify and reject background. The time-of-flight (ToF) between D1 and D2 was used to discriminate between downward-moving and upward-moving gamma rays, and also to distinguish between photons (moving at the speed of light) from slower moving neutrons (Fig. 4). (Neutrons generated by cosmic ray interactions in the atmosphere and in the spacecraft are a significant on-orbit background.) The 1σ ToF resolution was about 1 ns, which, given the 1.5 m separation of the D1 and D2 detector layers, was more than adequate to distinguish between upward and downward photons and also to identify a significant neutron component (Fig. 4). In addition

Journal of Physics: Conference Series **763** (2016) 012008 doi:10.1088/1742-6596/763/1/012008

to ToF, the D1 detectors also incorporated pulse shape measurements. The pulse shape could be used to distinguish true Compton scatter events from inelastic neutron scattering events(Fig. 5). This provided a second level of neutron rejection. Valid events could be identified utilizing both ToF and pulse-shape discrimination (PSD) measurements, along with the charged-particle rejection afforded by the plastic anti-coincidence system.

COMPTEL proved to be a successful project. It demonstrated techniques that could be used to limit the background, but the sensitivity was limited, in part because of its limited effective area. During its nine year mission, it detected only about 25 point-like sources on the entire sky [18], but it also proved its utility by mapping diffuse emissions in the galaxy, most notably the 1.8 MeV line from radioactive ^{26}Al [19]. Background rejection techniques were also key in the measurement of the isotropic cosmic diffuse γ-ray emission, showing that previous measurements had suffered from background-related issues.

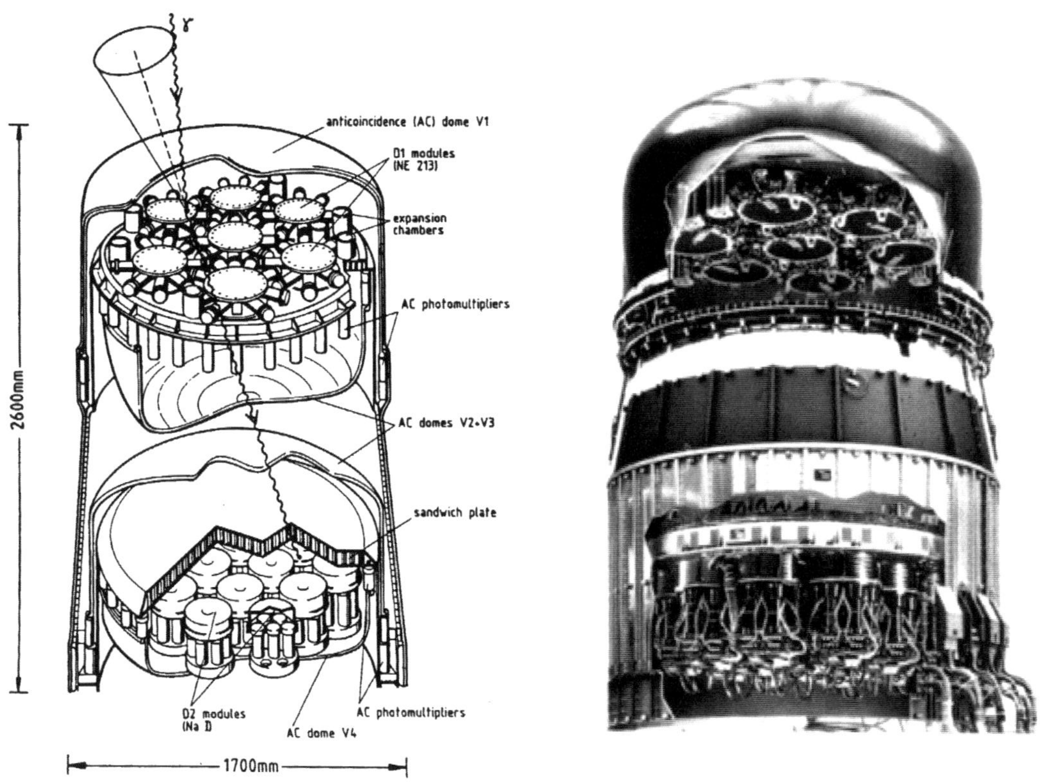

Figure 1. The COMPTEL instrument used Compton imaging for 1–30 MeV γ-rays, but had limited polarization sensitivity.

Figure 2. The D1 layer of COMPTEL consisted of seven large volume liquid scintillator detectors.

Figure 3. The D2 layer of COMPTEL consisted of fourteen large volume NaI(Tl) scintillator detectors (one of which had yet to be installed in the photo on the right).

4. An Advanced Compton Telescope

For several years, we have been investigating ways of improving on the COMPTEL design using the latest in scintillator technology [20, 21, 22]. This effort is embodied in an instrument that we call ASCOT (Advanced Scintillator COmpton Telescope) [23, 24, 25]. The ASCOT design incorporates the use of modern scintillator materials to retain the simple design and proven, ToF background rejection method, while using modern light readout devices to eliminate bulky and fragile photomultiplier tubes (PMTs) with their high-voltage requirements.

Recent advances in inorganic scintillator technology, such as the availability of Cerium-doped Lanthanum tri-Bromide ($LaBr_3$(Ce) [26]) and Cerium tri-Bromide ($CeBr_3$ [27]) have opened up a wide range of exciting new possibilities for high-energy astronomy and space science. Both materials offer good energy resolution for gamma-ray measurements, competitive with that of semiconductor detectors while retaining the many strengths of scintillators: simple and reliable implementation, high stopping power, large volume, room temperature operation, and very fast timing. In addition, new growth techniques have greatly increased the utility of organic

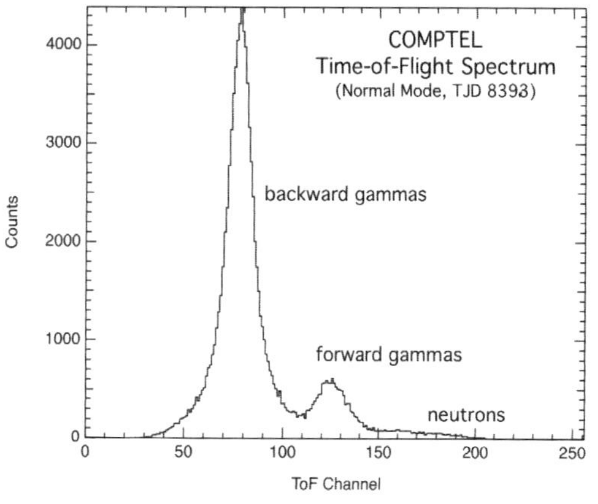

Figure 4. A typical COMPTEL ToF spectrum. The separation between forward and background gamma-rays is about 10 ns.

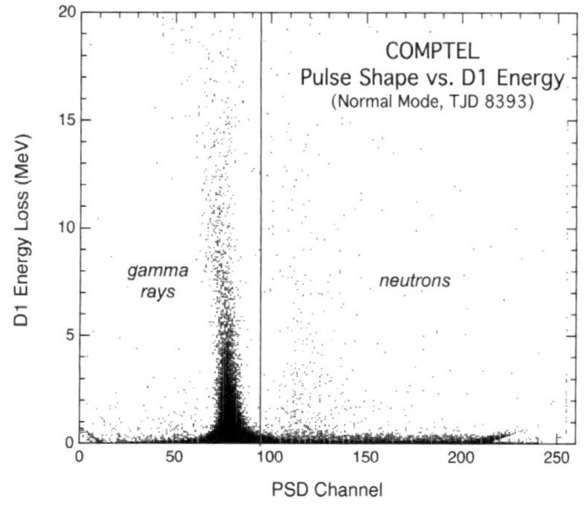

Figure 5. A typical COMPTEL PSD spectrum (vs D1 energy), showing the separation between gamma-rays and neutrons.

scintillator crystals, such as stilbene [27] and p-terphenyl [28], that feature high light output and good differentiation between electron and proton signals via PSD. Combined with recent advances in light readout technologies, such as Si photomultipliers (SiPMs), the scintillators can now be used in configurations with significantly reduced mass, volume, and power.

A balloon prototype of ASCOT is currently being prepared for a balloon flight from Ft. Sumner, NM in the fall of 2017 [29]. The instrument (Fig. 6) is composed of arrays of 15 mm $\times$ 15 mm $\times$ 25 mm scintillator elements. Each scintillator element is read out by a 2 $\times$ 2 sub-array of SensL SiPMs. The D1 detector assembly is composed of two layers of p-terphenyl scintillator (one module of which is shown in Fig. 7). The D2 detector assembly is composed of a single layer of $CeBr_3$ scintillator (one module of which is shown in Fig. 8). The energy resolution of one single $CeBr_3$ element has been measured as 6.1 % at 662 keV (FWHM). The ToF resolution of ASCOT has been measured using one p-terphenyl pixel and one $CeBr_3$ pixel. At 1.2 MeV, a 1σ ToF resolution of 225 ps was measured. This represents a four-fold improvement over the 1 ns resolution of COMPTEL. The improved ToF resolution enables a much smaller D1-D2 separation of 15 cm (as compared with the 150 cm of COMPTEL). The much smaller separation results in a significant improvement of sensitivity and a much larger FoV. It is hoped that the first results from ASCOT will be obtained next fall, demonstrating that a next-generation COMPTEL instrument can be be based on the same basic design, but with the latest in scintillator technologies.

5. Importance of Scintillators in Astrophysical Research

Scintillation detectors remain an important component of many γ-ray telescopes. They represent a reliable, flight-proven technology that is robust under a variety of operating conditions. They are relatively low in cost, and can be fabricated in a variety of shapes and sizes. New readout technology, such as SiPMs, make scintillators even more attractive by reducing mass and volume, while maintaining the desirable characteristics of PMTs.

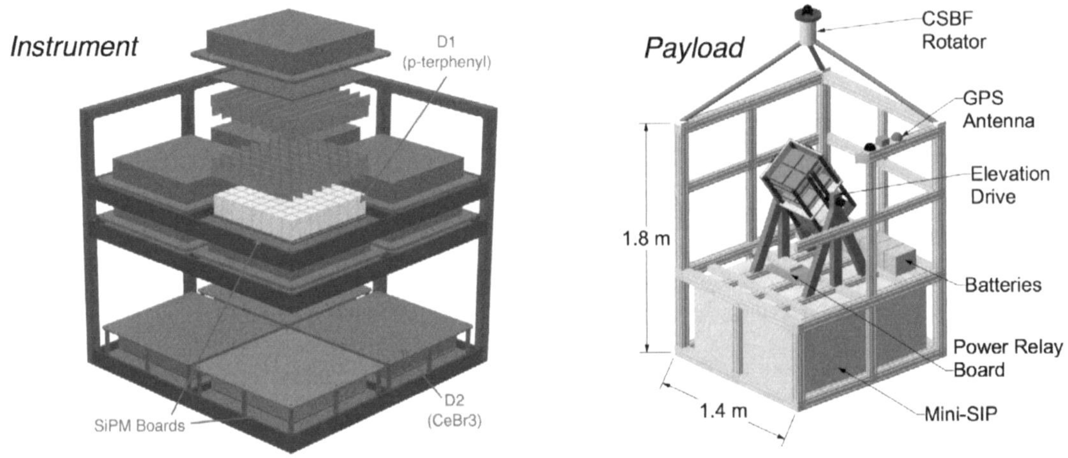

Figure 6. The design of the ASCOT balloon payload, which is scheduled for a balloon flight from Ft. Sumner, NM in the fall of 2017. The instrument itself is shown on the left. The full balloon payload is shown on the right.

Figure 7. One of eight p-terphenyl scintillator modules that will be used for the D1 layer of ASCOT.

Figure 8. One of four $CeBr_3$ scintillator modules that will be used for the D2 layer of ASCOT.

5.1. Acknowledgments
This work was supported by NASA Grants NNX15AI70G, NNX12AB36G, NNX13AC89G, NNX08AK47G, NNX10AC14G, and NNX10AG10G.

References

[1] Caroli E, Stephen J B, Di Cocco G, Natalucci L and Spizzichino A 1987 *Space Science Reviews* **45** 349–403
[2] Bouchet L, Roques J P, Ballet J, Goldwurm A and Paul J 2001 *Astrophysical Journal* **548** 990–1009
[3] Ubertini P, Lebrun F, Di Cocco G, Bazzano A, Bird A J, Broenstad K, Goldwurm A, La Rosa G, Labanti C, Laurent P, Mirabel I F, Quadrini E M, Ramsey B, Reglero V, Sabau L, Sacco B, Staubert R, Vigroux L, Weisskopf M C and Zdziarski A A 2003 *Astronomy and Astrophysics* **411** L131–L139
[4] Vedrenne G, Roques J P, Schönfelder V, Mandrou P, Lichti G G, von Kienlin A, Cordier B, Schanne S, Knödlseder J, Skinner G, Jean P, Sánchez F, Caraveo P A, Teegarden B, von Ballmoos P, Bouchet L, Paul P, Matteson J L, Boggs S E, Wunderer C, Leleux P, Weidenspointner G, Durouchoux P, Diehl R, Strong A W, Cassé M, Clair M A and André Y 2003 *Astronomy and Astrophysics* **411** L63–L70

[5] Barthelmy S D 2004 *Proceedings of SPIE* **5165** 175–189

[6] Schönfelder V, Aarts H J M, Bennett K, de Boer H, Clear J, Collmar W, Connors A, Deerenberg A J M, Diehl R, von Dordrecht A, den Herder J W, Hermsen W, Kippen R M, Kuiper L M, Lichti G G, Lockwood J A, Macri J R, McConnell M L, Morris D, Much R, Ryan J M, Simpson G, Snelling M, Stacy G, Steinle H, Strong A W, Swanenburg B N, Taylor B, de Vries C and Winkler C 1993 *Astrophysical Journal Supplement Series* **86** 657–692

[7] Bignami G F, Boella G, Burger J J, Taylor B G, Keirle P, Paul J A, Mayer-Hasselwander H A, Pfeffermann E, SCARSI L and Swanenburg B N 1975 *Space Science Instrumentation* **1** 245–268

[8] Thompson D J, Bertsch D L, Fichtel C E, Hartman R C, HOFSTADTER R, Hughes E B, Hunter S D, Hughlock B W, Kanbach G, Kniffen D A, Lin Y C, Mattox J R, Mayer-Hasselwander H A, von Montigny C, Nolan P L, Nel H I, Pinkau K, Rothermel H, Schneid E J, Sommer M, Sreekumar P, Tieger D and Walker A H 1993 *Astrophysical Journal Supplement Series* **86** 629–656

[9] Atwood W B, Abdo A A, Ackermann M, Althouse W, Anderson B, Axelsson M, Baldini L, Ballet J, Band D L, Barbiellini G, Bartelt J, Bastieri D, Baughman B M, Bechtol K, Bédérède D, Bellardi F, Bellazzini R, Berenji B, Bignami G F, Bisello D, Bissaldi E, Blandford R D, Bloom E D, Bogart J R, Bonamente E, Bonnell J, Borgland A W, Bouvier A, Bregeon J, Brez A, Brigida M, Bruel P, Burnett T H, Busetto G, Caliandro G A, Cameron R A, Caraveo P A, Carius S, Carlson P, Casandjian J M, Cavazzuti E, Ceccanti M, Cecchi C, Charles E, Chekhtman A, Cheung C C, Chiang J, Chipaux R, Cillis A N, Ciprini S, Claus R, Cohen-Tanugi J, Condamoor S, Conrad J, Corbet R, Corucci L, Costamante L, Cutini S, Davis D S, Decotigny D, DeKlotz M, Dermer C D, de Angelis A, Digel S W, Do Couto E Silva E, Drell P S, Dubois R, Dumora D, Edmonds Y, Fabiani D, Farnier C, Favuzzi C, Flath D L, Fleury P, Focke W B, Funk S, Fusco P, Gargano F, Gasparrini D, Gehrels N A, Gentit F X, Germani S, Giebels B, Giglietto N, Giommi P, Giordano F, Glanzman T, Godfrey G, Grenier I A, Grondin M H, Grove J E, Guillemot L, Guiriec S, Haller G, Harding A K, Hart P A, Hays E, Healey S E, Hirayama M, Hjalmarsdotter L, Horn R, Hughes R E, Jóhannesson G, Johansson G, Johnson A S, Johnson R P, Johnson T J, Johnson W N, Kamae T, Katagiri H, Kataoka J, Kavelaars A, Kawai N, Kelly H, Kerr M, Klamra W, Knödlseder J, Kocian M L, Komin N, Kuehn F, Kuss M, Landriu D, Latronico L, Lee B, Lee S H, Lemoine-Goumard M, Lionetto A M, Longo F, Loparco F, Lott B, Lovellette M N, Lubrano P, Madejski G M, Makeev A, Marangelli B, Massai M M, Mazziotta M N, McEnery J E, Menon N, Meurer C, Michelson P F, Minuti M, Mirizzi N, Mitthumsiri W, Mizuno T, Moiseev A A, Monte C, Monzani M E, Moretti E, Morselli A, Moskalenko I V, Murgia S, Nakamori T, Nishino S, Nolan P L, Norris J P, Nuss E, Ohno M, Ohsugi T, Omodei N, Orlando E, Ormes J F, Paccagnella A, Paneque D, Panetta J H, Parent D, Pearce M, Pepe M, Perazzo A, Pesce-Rollins M, Picozza P, Pieri L, Pinchera M, Piron F, Porter T A, Poupard L, Rainò S, Rando R, Rapposelli E, Razzano M, Reimer A, Reimer O, Reposeur T, Reyes L C, Ritz S, Rochester L S, Rodriguez A Y, Romani R W, Roth M, Russell J J, Ryde F, Sabatini S, Sadrozinski H F W, Sanchez D, Sander A, Sapozhnikov L, Parkinson P M S, Scargle J D, Schalk T L, Scolieri G, Sgro C, Share G H, Shaw M, Shimokawabe T, Shrader C, Sierpowska-Bartosik A, Siskind E J, Smith D A, Smith P D, Spandre G, Spinelli P, Starck J L, Stephens T E, Strickman M S, Strong A W, Suson D J, Tajima H, Takahashi H, Takahashi T, Tanaka T, Tenze A, Tether S, Thayer J B, Thayer J G, Thompson D J, Tibaldo L, Tibolla O, Torres D F, Tosti G, Tramacere A, Turri M, Usher T L, Vilchez N, Vitale V, Wang P, Watters K, Winer B L, Wood K S, Ylinen T and Ziegler M 2009 *Astrophysical Journal* **697** 1071–1102

[10] Tavani M, Barbiellini G, Argan A, Boffelli F, Bulgarelli A, Caraveo P A, Cattaneo P W, Chen A W, Cocco V, Costa E, D'Ammando F, del Monte E, de Paris G, Di Cocco G, Di Persio G, Donnarumma I, Evangelista Y, Feroci M, Ferrari A, Fiorini M, Fornari F, Fuschino F, Froysland T, Frutti M, Galli M, Gianotti F, Giuliani A, Labanti C, Lapshov I, Lazzarotto F, Liello F, Lipari P, Longo F, Mattaini E, Marisaldi M, Mastropietro M, Mauri A, Mauri F, Mereghetti S, Morelli E, Morselli A, Pacciani L, Pellizzoni A, Perotti F, Piano G, Picozza P, Pontoni C, Porrovecchio G, Prest M, Pucella G, Rapisarda M, Rappoldi A, Rossi E, Rubini A, Soffitta P, Traci A, Trifoglio M, Trois A, Vallazza E, Vercellone S, Vittorini V, Zambra A, Zanello D, Pittori C, Preger B, Santolamazza P, Verrecchia F, Giommi P, Colafrancesco S, Antonelli A, Cutini S, Gasparrini D, Stellato S, Fanari G, Primavera R, Tamburelli F, Viola F, Guarrera G, Salotti L, D'Amico F, Marchetti E, Crisconio M, Sabatini P, Annoni G, Alia S, Longoni A, Sanquerin R, Battilana M, Concari P, Dessimone E, Grossi R, Parise A, Monzani F, Artina E, Pavesi R, Marseguerra G, Nicolini L, Scandelli L, Soli L, Vettorello V, Zardetto E, Bonati A, Maltecca L, D'Alba E, Patané M, Babini G, Onorati F, Acquaroli L, Angelucci M, Morelli B, Agostara C, Cerone M, Michetti A, Tempesta P, D'Eramo S, Rocca F, Giannini F, Borghi G, Garavelli B, Conte M, Balasini M, Ferrario I, Vanotti M, Collavo E and Giacomazzo M 2009 *Astronomy and Astrophysics* **502** 995–1013

[11] Gehrels N A, Fichtel C E, Fishman G J, Kurfess J D and Schönfelder V 1993 *Scientific American* **269** 38

[12] Winkler C, Courvoisier T J L, Di Cocco G, Gehrels N A, Giménez A, Grebenev S, Hermsen W, Mas-Hesse J M, Lebrun F, Lund N, Palumbo G G C, Paul J, Roques J P, Schnopper H, Schönfelder V, Sunyaev R,

Teegarden B, Ubertini P, Vedrenne G and Dean A J J 2003 *Astronomy and Astrophysics* **411** L1–L6

[13] Boggs S E 2006 *New Astronomy Reviews* **50** 604–607

[14] Boggs S E, Coburn W, Smith D M, Bowen J D, Jean P, Kregenow J M, Lin R P and von Ballmoos P 2004 *New Astronomy Reviews* **48** 251–255

[15] Novikova E I, Wulf E A, Phlips B F, Kurfess J D, Zoglauer A, Weidenspointner G and Kippen R M 2005 Simulations of a Si-based Advanced Compton Telescope *Nuclear Science Symposium Conference Record, 2005 IEEE* pp 985–989

[16] Aprile E, Curioni A, Giboni K L, Kobayashi M, Oberlack U G and Zhang S 2008 *Nuclear Instruments and Methods in Physics Research Section A* **593** 414–425

[17] Aprile E, Curioni A, Egorova V, Giboni K, Oberlack U G, Ventura S, Doke T, Takizawa K, Chupp E L and Dunphy P 2001 A liquid xenon time projection chamber for gamma-ray imaging in astrophysics: present status and future directions *Nuclear Instruments & Methods in Physics Research Section a-Accelerators Spectrometers Detectors and Associated Equipment* (Columbia Univ, Dept Phys, Columbia Astrophys Lab, New York, NY 10027 USA) pp 256–261

[18] Schönfelder V, Bennett K, Blom J J, Bloemen H, Collmar W, Connors A, Diehl R, Hermsen W, Iyudin A, Kippen R M, Knödlseder J, Kuiper L M, Lichti G G, McConnell M L, Morris D, Much R, Oberlack U G, Ryan J M, Stacy G, Steinle H, Strong A W, Suleiman R, van Dijk R, Varendorff M, Winkler C and Williams O R 2000 *Astronomy and Astrophysics Supplement Series* **143** 145–179

[19] Diehl R 1995 *Experimental Astronomy* **6** 103–108

[20] Bloser P F, Legere J S, Bancroft C M, Jablonski L F, Wurtz J R, Ertley C D, McConnell M L and Ryan J M 2014 *Nuclear Inst. and Methods in Physics Research* **763** 26–35

[21] Bloser P F, Legere J S, Bancroft C M, McConnell M L and Ryan J M 2014 Scintillators with silicon photomultiplier readouts for high-energy astrophysics and heliophysics *Proceedings of the SPIE* ed Takahashi T, den Herder J W A and Bautz M Space Science Ctr., The Univ. of New Hampshire (United States) (SPIE) pp 914414–914414–11

[22] Bloser P F, Legere J S, Bancroft C M, McConnell M L, Ryan J M and Schwadron N 2013 Scintillator gamma-ray detectors with silicon photomultiplier readouts for high-energy astronomy *Proceedings of the SPIE* ed Siegmund O H The Univ. of New Hampshire (United States) (SPIE) pp 88590A–88590A–8

[23] Bloser P F, Legere J S, Bancroft C M, Ryan J M and McConnell M L 2016 *Nuclear Inst. and Methods in Physics Research* **812** 92–103

[24] Bloser P F, Ryan J M, Legere J S, Julien M, Bancroft C M, McConnell M L, Wallace M, Kippen R M and Tornga S 2010 *Proceedings of SPIE* **7732** 64

[25] Bloser P F, Ryan J M, Legere J S, Julien M, Bancroft C M, McConnell M L, Wallace M, Kippen R M and Tornga S 2009 *Proceedings of SPIE* **7435** 12

[26] van Loef E V D, Dorenbos P, van Eijk C W E, Krämer K and Güdel H U 2001 *Applied Physics Letters* **79** 1573–1575

[27] Kim H D, Cho G S and Kim H J 2013 *Radiation Measurements* **58** 133–137

[28] Matei C, Hambsch F J and Oberstedt S 2012 *Nuclear Instruments and Methods in Physics Research Section A* **676** 135–139

[29] Bloser P F, Sharma T, Legere J S, Bancroft C M, McConnell M L, Ryan J M and Wright A M 2016 *Proceedings SPIE* **9905** 99056K–99056K–7

Journal of Physics: Conference Series **763** (2016) 012009 doi:10.1088/1742-6596/763/1/012009

The XGS instrument on-board THESEUS

F. Fuschino[a,b,c], R. Campana[b,c], C. Labanti[b,c], M. Marisaldi[b,c], L. Amati[b], M. Fiorini[d], M. Uslenghi[d], G. Baldazzi[a,c], Y. Evangelista[e], I. Elmi[f], M. Feroci[e], F. Frontera[b,g], A. Rachevski[h], L. P. Rignanese[a,c], A. Vacchi[i,h], G. Zampa[h], N. Zampa[h], I. Rashevskaya[l], P. Bellutti[m] and C. Piemonte[m]

[a] Department of Physics, University of Bologna, Viale Berti Pichat 6/2, I-40127 Bologna, Italy
[b] INAF/IASF-Bologna, Via Gobetti 101, I-40129 Bologna, Italy
[c] INFN - Sezione di Bologna, Viale Berti Pichat 6/2, I-40127 Bologna, Italy
[d] INAF/IASF-Milano, Via Bassini 23, I-20133 Milano, Italy
[e] INAF/IAPS, Via Fosso del Cavaliere 100, I-00133 Roma, Italy
[f] CNR/IMM, Via Gobetti 101, I-40129 Bologna, Italy
[g] Department of Physics, University of Ferrara, Via Giuseppe Saragat 1, I-44122 Ferrara, Italy
[h] INFN - Sezione di Trieste, Padriciano 99, I-34127 Trieste, Italy
[i] Mathematics and Informatics Dept., Udine University, Via delle Scienze 206, I-33100 Udine, Italy
[l] TIFPA Dept., Via Sommarive, 14, 38123 Povo TN, Italy
[m] CMM Dept., Burno Kessler Foundation (FBK), Via Sommarive 18, I-38123 Povo, Italy

E-mail: fuschino@iasfbo.inaf.it

Abstract. Consolidated techniques used for space-borne X-ray and gamma-ray instruments are based on the use of scintillators coupled to Silicon photo-detectors. This technology associated with modern very low noise read-out electronics allows the design of innovative architectures able to reduce drastically the system complexity and power consumption, also with a moderate-to-high number of channels. These detector architectures can be exploited in the design of space instrumentation for gamma-spectroscopy with the benefit of possible smart background rejection strategies. We describe a detector prototype with 3D imaging capabilities to be employed in future gamma-ray and particle space missions in the 0.002–100 MeV energy range. The instrument is based on a stack of scintillating bars read out by Silicon Drift Detectors (SDDs) at both ends. The spatial segmentation and the crystal double-side readout allow a 3D position reconstruction with ~3 mm accuracy within the full active volume, using a 2D readout along the two external faces of the detector. Furthermore, one of the side of SDDs can be used simultaneously to detect X-rays in the 2–30 keV energy range. The characteristics of this instrument make it suitable in next generation gamma-ray and particle space missions for Earth or outer space observations, and it will be briefly illustrated.

1. Introduction

Gamma-Ray Bursts (GRBs) are one of the most intriguing and challenging phenomena for modern science. Because of their luminosities up to more than 10^{52} erg/s, their redshift distribution extending up to $z > 9$ and their association with peculiar core-collapse supernovae and with neutron star/black-hole mergers, their study is of very high interest for several fields of astrophysics. Despite the huge

observational advances in the last 15 years, several open issues still affect our comprehension of these phenomena, and their exploitation for fundamental physics and cosmology.
To address these fundamental issues, time resolved spectroscopy (and possibly polarimetry) of the GRB prompt emission over a broad energy range from few keV to several MeV is needed. A low energy boundary (below a few keV) is useful to investigate possible interactions of the prompt emission with the circumburst medium, while a continuous energy coverage in gamma-rays allows to measure a broad range of burst peak energies. We are therefore working towards a prototype (X and Gamma Spectrometer, XGS) for a monolithic system, which allows spectroscopy and timing of high-energy transient events, over an unprecedented broad energy band. The proposed detector combines a low energy threshold (~2 keV) and energy resolution significantly better than that of any other GRB detection system based on scintillators (e.g., Fermi/GBM) up to several MeV or CZT/CdTe semiconductors (INTEGRAL/ISGRI, Swift/BAT) up to 30–50 keV. The timing capability is of few micro-seconds resolution over the whole energy band, and a possible polarimetric capabilities is linked to the true three-dimensional photon interaction reconstruction.
In this paper we report on the progress of this currently undergoing prototype activity, in preparation for the proposed XGS instrument on board the THESEUS space mission [1][2].

2. The XGS instrument architecture

Aiming at designing a compact instrument with a very large sensitivity band, the XGS instrument combines both detection capabilities of silicon detector: direct X-ray detection in the energy range from ~2 keV up to 30–50 keV, and detection of scintillating light from inorganic scintillators in the energy range from 20 keV up to several MeV [3][4].

When the scintillator is shaped like a bar and two Silicon devices are coupled at the ends, both interacting position and overall energy released in the bar can be derived combining the response of two photo-detectors. In this concept (Figure 1), the Silicon detectors on the "top" side of the bars play the double role of scintillator read-out and of independent X-ray detector. The inorganic scintillator bar, made of CsI(Tl), is coupled at both ends with a Silicon Drift Detector [5] developed by INFN-Trieste and Fondazione Bruno Kessler (FBK, Trento) within the framework of the ReDSoX collaboration.

Profiting of the good spectroscopic performance of SDDs, when coupled with low noise Front End Electronics (FEE) [6], the sensitivity band of the proposed architecture is therefore from ~2 to 50 keV (X-events) for silicon and from 20 keV to several MeVs (G-events) for CsI(Tl), allowing for some overlap of the two operating modes, as shown in Figure 1.

The X-rays and γ-rays, detected in the Silicon and in the scintillator respectively, can be distinguished by pulse shape discrimination techniques, through the different risetimes of the corresponding preamplifier signal. In the case of "X-events", the risetime is dominated by the anode collection time (~100 ns), while for "G-events" the signal rises following the characteristic CsI(Tl) scintillation time constants and different light paths, amounting to a few μs.

A CAD sketch of the instrument concept is shown in Figure 2. Four scintillator bars (each about $4.5\times4.5\times45$ mm^3) are coupled with an array of 2×2 single-cell SDDs, with an active area of 25 mm^2 each. This constitutes the fundamental unit of the instrument module. To restrict the field of view to about 0.15 sr, increasing also the sensitivity of the instrument, a passive collimator can be placed on the top of the bar/detectors assembly.

3. Preliminary prototype

In order to demonstrate the performances of the proposed instrument, a prototype is being developed in the framework of an INAF-funded project. The overall setup is shown in Figure 3. The front-end electronics is composed of a discrete-elements charge sensitive preamplifier designed at INAF/IASF (PA-001), operated in continuous reset configuration. In order to guarantee good performances, the input stage of the charge preamplifier is arranged near the anode. The same PCB board house four 2×2 single-cell SDD arrays, for a total of 16 channels (Figure 3). The complete detector combines four such modules for each side of the bars. An additional plug-in motherboard holds single channel preamplifiers. The preamplifier output is finally acquired trough a fast digitizer and then digitally processed.

Journal of Physics: Conference Series **763** (2016) 012009 doi:10.1088/1742-6596/763/1/012009

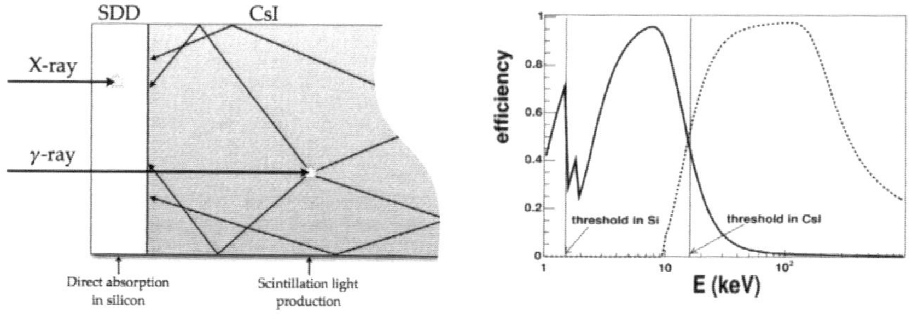

Figure 1. LEFT: The detector concept. Low energy X-rays are absorbed directly in the silicon bulk of the SDD, while high energy X-rays produce scintillation light in the CsI(Tl) crystal, that is collected by the SDDs. RIGHT: When SDD is used the detection efficiency can be extended down to few keV, reducing also the low threshold for scintillator and overlapping the two operating modes.

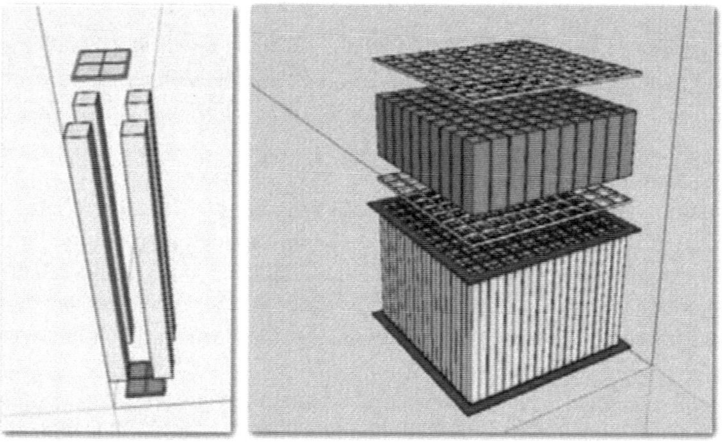

Figure 2. Left panel: the fundamental unit of XGS, composed of 4 scintillator bars coupled to two 2×2 SDD arrays. Right panel: the complete XGS module (CAD sketches).

The choice of a digital signal processing was justified by its large flexibility in this development phase, since it allows simultaneous testing of different shaping algorithms, to characterize the noise performance of the system in a rapid and extremely flexible way. Of course a custom-designed analog ASIC will be mandatory for an actual space-ready instrument, and the present prototype will allow to drive the design of this front-end electronics component.

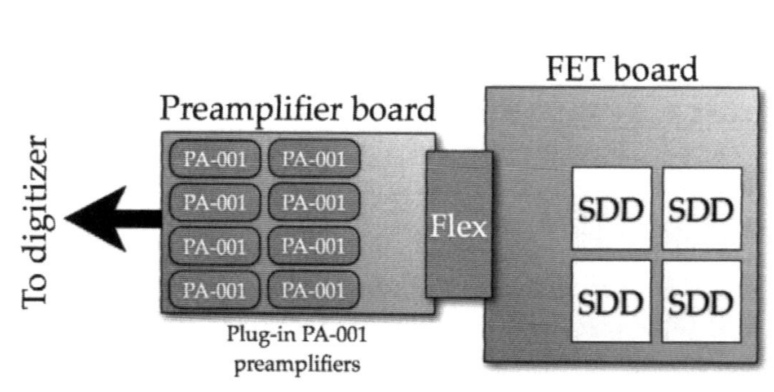

Figure 3. The overall setup of the XGS prototype

In this activity the digitized waveform was then processed with a digital trapezoidal filter [7]. When the trapezoidal filter with symmetric rise and fall times and a flat-top time is selected, it can be easily implemented as a cascade of finite and infinite impulse response filters, and is a good approximation of the "ideal" filter. The trapezoid was then sampled around the middle of the flat-top to extract the pulse height, proportional to the input energy.

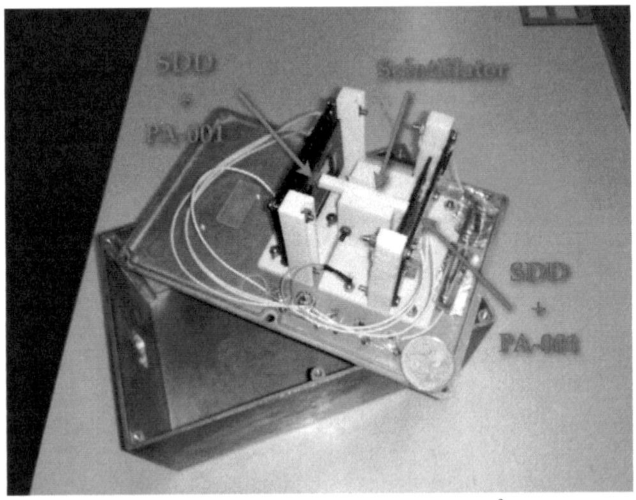

Figure 4. Preliminary XGS prototype. A single 4.5×4.5×45 mm^3 scintillator bar is coupled at both ends with 5×5 cm^2 single-cell SDDs. Each SDD is then coupled with a discrete components charge sensitive preamplifier.

In order to investigate the optical coupling between the scintillator bar and the SDDs, and to derive realistic performance figures for the complete XGS prototype, a single-bar, 2-channel sub-unit, was realized and characterized (Figure 4). The scintillator bar shape is the same envisaged for XGS: 4.5×4.5×45 mm^3. In Figure 5 is shown a simultaneous acquisition at room temperature of two radioactive sources: ^{137}Cs and ^{241}Am. The digitizer acquired both types of events simultaneously and then by means pulse shape discrimination techniques X-events and G-events were distinguished. Two optimized digital filters were then applied separately for the two categories. In this case, an energy resolution of about 5.8% on the 662 keV line is obtained when the system is operated at room temperature. The good discrimination capabilities of the digital method were demonstrated, showing also a comparable result with respect to a more traditional analog spectroscopic chain coupled to a commercial multichannel analyzer.

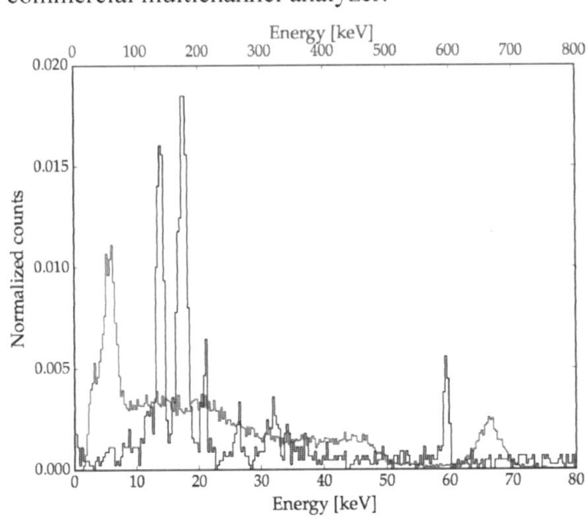

Figure 5. Simultaneous acquisition of a ^{137}Cs and ^{241}Am radioactive sources. The black curve is the spectrum of X-events, while the blue one is the spectrum of G-events. Both types of events were acquired with a digital backend electronics and digitally shaped. An energy resolution of about 5.8% on the 662 keV line is obtained when the system is operated at room temperature.

Measurements of the light attenuation, spatial and energetic resolution were performed using a collimated (~2 mm spot) ^{137}Cs radioactive source at various positions along the bar. The signal collected at each SDD can be represented, in a simplified model [8][9], as $U(x) \quad E\, u_0\, e^{-\alpha(\frac{L}{2}-x)}$; where L is the bar length, x is the position along the bar, α is the light attenuation coefficient, E the input energy and u_0 is the signal output at the bar edge (that includes the light yield of the scintillator and the quality of the optical coupling). The energy and the interaction position can be derived from the signals U_A, U_B collected by both detectors as $E \propto \sqrt{\frac{U_A}{U_B}}$ and $x \propto \log\frac{U_A}{U_B}$.

In Figure 6, left panel, the light attenuation measured by both SDDs is shown. An attenuation coefficient of ~0.021 cm^{-1}, consistent with [9][10], and an overall light output of ~27 e^{-}/keV were measured. In the right panel of Figure 6 the position reconstruction along the bar is shown. The average FWHM of the reconstructed position is ~3 mm. Considering the source spot size (~2 mm), the derived intrinsic resolution in position is around 2.2 mm.

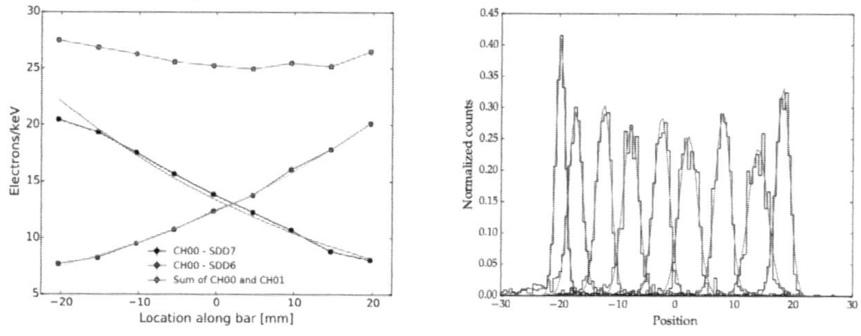

Figure 6. LEFT: Light attenuation along the bar. The signal from both detectors is shown. RIGHT: Reconstructed position for various measurements along the bar.

In Figure 7 a spectrum obtained with the collimated source at the bar center is shown. An energy resolution of 4.9% at 662 keV and a resolution of ~41.8% at 32 keV were measured, when the system was operated at -20 °C. These results are compatible with a lower threshold of about 20 keV and confirm the expected performance of the proposed system.

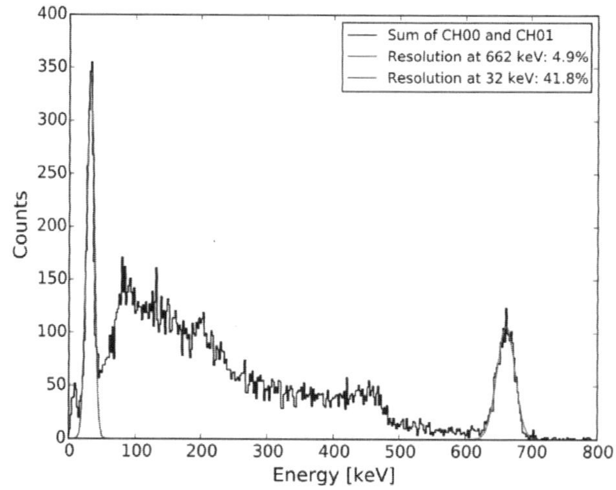

Figure 7. Spectrum of a collimated ^{137}Cs radioactive source at the center of the bar. The spectrum is reconstructed combining both SDDs signals. Gaussian fits of the 662 keV and 32 keV lines are also shown.

4. Back-end Architecture

For the complete XGS prototype, a digital back-end architecture is required. The attractive low-cost solution proposed is based on the commercial ADC board developed by RedPitaya. This board is based on a Xilinx Zynq 7010 system-on-chip, in which a FPGA is coupled to an ARM Cortex 9 CPU, allowing a great flexibility in the deployment and configuration. Each board, composed by 14-bit, 125 MS/s dual channel ADC, is equipped with an external suitable trigger logic and a custom software developed at IASF-Bologna in Python using the Qt libraries. The GUI allows to configure several RedPitaya boards, handling the acquisition and quicklook operations.

5. Conclusions

The preliminary prototypes developed (Section 3) have shown the good overall performance of the system. We are currently working towards the implementation of the complete XGS prototype, using an innovative, low-cost digital back-end (Section 4). The aim is to demonstrate the feasibility of this compact and modular architecture concept (and to drive the development of a dedicated analog front-end electronic circuit) for a space-borne, wide energy range, sensitive photon detector for the future missions for high energy astrophysics.

Acknowledgments

This project is funded by INAF through a TecnoPRIN 2014 grant. The Silicon Drift Detectors have been developed in the framework of the ReDSoX collaboration and FBK/INFN agreement.

References

[1] W., Yuan, L. Amati, J.K. Cannizzo, B. Cordier, N. Gehrels, G. Ghirlanda, D. Götz, N. Produit, Y. Qiu, J. Sun, N.R. Tanvir, J. Wei, C. Zhang, 2016, "Perspectives on Gamma-Ray Burst Physics and Cosmology with Next Generation Facilities", Space Science Reviews, in press

[2] L. Amati, P. O'Brien, D. Gotz, "The Transient High Energy Sky and Early Universe Surveyor (THESEUS)" Proceedings of SPIE, DOI: http://dx.doi.org/10.1117/12.2231525

[3] Marisaldi, M., Labanti, C., and Soltau, H., "A Pulse Shape Discrimination Gamma-Ray Detector Based on a Silicon Drift Chamber Coupled to a CsI(Tl) Scintillator: Prospects for a 1 keV 1 MeV Monolithic Detector," IEEE Transactions on Nuclear Science 51, 1916–1922 (2004).

[4] Marisaldi, M., Labanti, C., Soltau, H., Fiorini, C., Longoni, A., and Perotti, F., "X- and Gamma-Ray Detection With a Silicon Drift Detector Coupled to a CsI(Tl) Scintillator Operated With Pulse Shape Dis- crimination Technique," IEEE Transactions on Nuclear Science 52, 1842–1848 (2005).

[5] Gatti, E. and Rehak, P., "Semiconductor drift chamber an application of a novel charge transport scheme," Nuclear Instruments and Methods in Physics Research 225(3), 608 – 614 (1984).

[6] Bertuccio, G., et al., "X-Ray Silicon Drift Detector–CMOS Front-End System with High Energy Resolution at Room Temperature", IEEE TNS, VOL. 63, (2016)

[7] Guzik, Z. and Krakowski, T., "Algorithms for digital -ray spectroscopy," Nukleonika 58, 333 (2013).

[8] Labanti, C., Caroli, E., Rossi, E., and Spizzichino, A., "Methods for optimizing the performances of position- sensitive CsI(Tl) scintillation bars with photodiode readout," Nuclear Instruments and Methods in Physics Research A 310, 327–331 (1991).

[9] Labanti, C., Marisaldi, M., Fuschino, F., Bastia, P., Negri, B., Perotti, F., and Soltau, H., "Position sensitive X and gamma-ray scintillator detector for new space telescopes," Proceedings of SPIE 7021, 702116 (2008).

[10] Marisaldi, M., Fiorini, C., Labanti, C., Longoni, A., Perotti, F., Rossi, E., and Soltau, H., "Silicon drift detectors coupled to CsI(Tl) scintillators for spaceborne gamma-ray detectors," Nuclear Physics B Proceedings Supplements 150, 190–193 (2006).

A comparison of emerging gamma detector technologies for airborne radiation monitoring

S J Bell[1], P Aitken-Smith[1], S Beeke[1,2], S M Collins[1], P H Regan[1,2] and R Shearman[1,2]

[1]AIR Division, National Physical Laboratory, Hampton Road, Teddington, TW11 0LW, UK
[2]Department of Physics, University of Surrey, Guildford, GU2 7XH, UK

Email: steven.bell@npl.co.uk

Abstract. This paper presents a comparison of new and emerging gamma detector technologies that have the potential to improve in-situ dose and radioactivity-in-air measurements for national monitoring networks. Five detectors were chosen for investigation; $LaBr_3$(Ce), $CeBr_3$, SiPM-CsI(Tl), Cd(Zn)Te and electromechanically-cooled HPGe. These detectors represent the full range of the price-performance matrix. Comparisons have been made of energy resolution, detection efficiency and minimum detectable activity by exposing each detector to a mixed radionuclide source drop-deposited across a filter. Other factors, such as internal radioactivity, linearity, size and cost have also been considered.

1. Introduction

Article 35 of the EURATOM treaty mandates continuous environmental monitoring of radioactivity-in-air [1]. The Chernobyl accident led to increased verification of treaty compliance and wide investment in radiation monitoring networks across Europe. These networks use gamma detectors to measure gamma dose-in-air and gamma emitting airborne particulates collected on filters. In the event of a radiological incident, the measurements made by these networks must be rapidly collected, analysed and assessed. The results are then forwarded to national governments where decisions with far reaching consequences may have to be made. For this reason it is critical that the gamma measurements are as accurate and complete as possible.

The majority of in-field gamma detectors currently used represent tried and tested but dated technology, such as Geiger Müller counters and NaI(Tl) spectrometers. The disaster at Fukushima has intensified research and investment in environmental monitoring and many European authorities are looking to upgrade their gamma radiation monitoring networks. New portable radioactivity-in-air monitoring systems developed by IJS (Slovenia) [2] and CIEMAT (Spain) [3] are examples of this new technology.

Presented is a comparison of new and emerging gamma detector technologies that have the potential to improve in-field dose and radioactivity-in-air measurements for national monitoring networks. The detectors investigated include $LaBr_3$(Ce), $CeBr_3$, SiPM-CsI(Tl), Cd(Zn)Te and electromechanically-cooled HPGe. These detectors cover the full price-performance matrix. The detectors have been compared in terms of energy resolution, detection efficiency, linearity, internal radioactivity and minimum detectable activity (MDA). These properties are of vital importance when choosing a suitable detector for in-situ gamma spectrometry of filters.

This work is part of the Euramet EMRP MetroERM project; a project to harmonise the measurement and analysis of airborne radiation following a radiological event [4].

2. Method

The experimental arrangement was chosen to mimic the deployment of a gamma spectrometer in-situ with a high-volume air sampler. It was envisaged that the air sampler would cease pumping after a given time and then the gamma measurement would commence. To achieve this, a spiked filter was prepared and placed 3 cm from the centre of the detector front window.

Measurements of the filter were made within a Cu/Cd lined Pb coffin, and taken over 24 hr. Full energy peak centroids and net areas were extracted from the spectra in order to perform energy and efficiency calibrations. The resolution, efficiency and linearity were assessed. Background measurements were completed over 24 hr to determine MDA values, according to the Currie method [5]. The calculated MDA values represented the "best possible scenario" as Cu/Cd/Pb shielding is unlikely to be available for deployable air monitors.

Details of the five gamma spectrometers investigated are given in section 2.1. Details of the filters prepared for the experiment are given in section 2.2.

2.1. Detectors

The five gamma spectrometers selected for comparison are listed in Table 1. The HPGe and Cd(Zn)Te detectors were self-contained units with integrated high voltage (HV), pre-amplifier, shaping amplifier, analogue-to-digital convertor (ADC) and multi-channel analyser (MCA). The CsI(Tl) was coupled to a silicon photomultiplier (SiPM) and was also mounted in a self-contained unit with integrated readout electronics. The amplifier gain and shaping time settings for the HPGe, Cd(Zn)Te and CsI(Tl) detectors were optimised by the manufacturers and not adjusted. The $LaBr_3$(Ce) [6] and $CeBr_3$ detectors were coupled to Hamamatsu R9779 PMTs and were read out through an Ortec 672 shaping amplifier and an Ortec 926 MCA. A discreet pre-amplifier was not used. The shaping time and gain were optimised to improve energy resolution. Data acquisition was controlled with MAESTRO or KSPECT depending on the detector being tested. Acquired spectra were exported to Excel and processed with a custom script to extract peak centroids and net counts to ensure continuity of spectral analysis.

Table 1. Details of the five gamma spectrometers investigated.

Detector	Size (cm)	Volume (cm^3)	Notes
HPGe	8.5 x 3.0	170	Electro-mechanically cooled Ortec Detective200 with p-type coaxial crystal
$CeBr_3$	3.8 x 5.1	58	Saint Gobain crystals coupled to
$LaBr_3$(Ce)	3.8 x 5.1	58	Hamamatsu R9779 PMTs [6]
Cd(Zn)Te	1.0 x 1.0 x 1.0	1	USB powered Kromek GR1 with co-planar grid electrodes
CsI(Tl)	2.5 x 2.5 x 5.1	33	USB powered Kromek Sigma50 with SiPM

2.2. Filters

The 20.5×10.5 cm^2 filter was prepared by drop-deposition of a mixed radionuclide solution. The solution was deposited evenly across a 1×1 cm^2 grid. The 200 point grid was divided into eight subsections and the dispensing bottle weighed after completion of each subsection to ensure uniformity. The maximum deviation of mass deposited to each subsection was no more than ± 0.7% from the average across the eight subsections. A total of just under 3 g of solution was deposited, providing a total activity of 29.11 ± 0.34 kBq. A photo of the filter is shown in figure 1. The activity concentration of the filter was chosen to mimic the activity collected on a filter being pumped at 1,000

m^3/hr for 48 hr in an atmosphere of 0.03 Bq/m^3 of ^{137}Cs; the minimum reporting level according to the EURATOM treaty [7]. The radionuclide content of the filter is given in table 2.

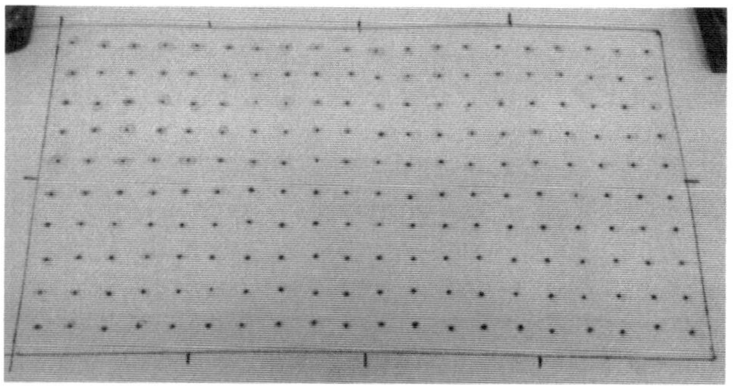

Figure 1. Photograph of 20.5 × 10.5 cm^2 "half-RASA" filter with 1 x 1 cm^2 grid. The mixed radionuclide solution was drop-deposited evenly across the 200 grid points.

Table 2. Details of the mixed radionuclide source drop-deposited on the filter. Uncertainties quoted with k = 2 coverage.

Radionuclide	Characteristic Gamma-ray Decay Energy (keV)	Activity (kBq)
^{241}Am	59.5	1.968 ± 0.028
^{109}Cd	88.0	8.12 ± 0.31
^{57}Co	122.1	0.4024 ± 0.010
^{139}Ce	165.9	0.3308 ± 0.0067
^{51}Cr	320.1	6.04 ± 0.11
^{113}Sn	391.7	1.191 ± 0.021
^{85}Sr	514.0	1.113 ± 0.020
^{137}Cs	661.7	1.472 ± 0.026
^{54}Mn	834.8	1.452 ± 0.019
^{88}Y	898.0 1836.1	2.368 ± 0.031
^{65}Zn	1115.5	2.990 ± 0.053
^{60}Co	1173.2 1332.5	1.671 ± 0.020

3. Results

3.1. Energy spectra, resolution and efficiency

Early warning air monitors should provide rapid and accurate information in the event of an incident. Fine energy resolution is vital for accurate radionuclide identification, which can assist in determining the source of radiation. The efficiency of a detector largely determines the speed with which radionuclide identification can be performed.

Significant differences in terms of energy resolution and detection efficiency were observed across the five detectors. This is illustrated in the comparison of energy spectra in figure 2. As expected, the HPGe detector produced the best FWHM energy resolution; 0.25% at 662 keV, followed by the Cd(Zn)Te detector; 1.50% at 662 keV. The energy resolution of the $LaBr_3$(Ce) detector was 3.78%; marginally better than that of the $CeBr_3$, which was 5.0%. The CsI(Tl) detector produced the poorest resolution, for example, failing to separate the 1115.5 keV ^{65}Zn peak from the 1173.2 keV ^{60}Co peak. A full resolution comparison is presented in figure 3.

Benefitting from a large volume and surface area, the HPGe detector produced the best detection efficiency. However the higher effective atomic numbers of the other detector materials went some way to compensate. This is demonstrated in table 3 where the efficiency and volume of each detector relative to that of HPGe is compared. For in-situ filter measurements, a compact detector is a necessity to avoid disruption of the air flow.

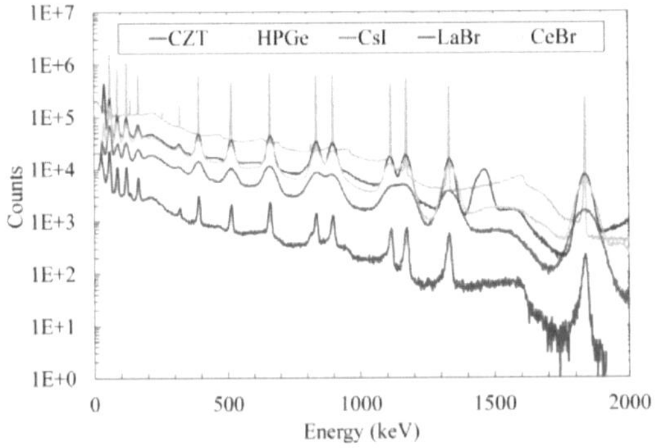

Figure 2. Comparison of energy spectra measured by detectors exposed to 29 kBq mixed radionuclide filter source for 24 hr.

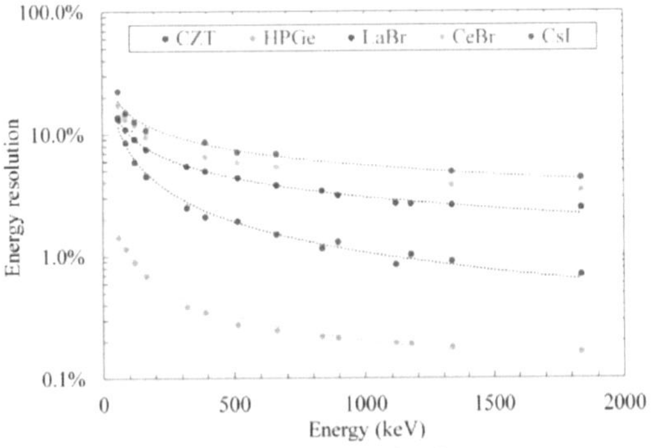

Figure 3. Comparison of detector energy resolution from 60 keV to 1.8 MeV. Each dataset has been fitted with a power function.

Table 3. Efficiency and relative volume of each detector. Uncertainties quoted with $k = 2$ coverage.

Detector	Absolute efficiency at 662 keV	Efficiency relative to HPGe	Volume relative to HPGe
HPGe	1.161(50)%	[n/a]	[n/a]
$CeBr_3$	0.649(11)%	56%	34%
$LaBr_3$(Ce)	0.658(11)%	56%	34%
Cd(Zn)Te	0.01167(28)%	1.0%	0.59%
CsI(Tl)	0.3116(61)%	27%	19%

3.2. Linearity

Knowledge of the linearity of a detector allows the user to determine the most appropriate energy calibration to perform. The linearity of each detector was assessed by comparing how the proportionality (channels per keV) varied across the energy range. The plot in figure 4 demonstrates the non-linearity of the $LaBr_3$(Ce), $CeBr_3$ and CsI(Tl) detectors, compared to the highly linear HPGe detector. A high order polynomial function was required to produce a satisfactory calibration of these detectors.

It was observed that the primary source of non-linearity for the $LaBr_3$(Ce)/$CeBr_3$ detectors was the lack of magnetic shield on the PMT. A comparison with a shielded $LaBr_3$(Ce) is shown in figure 4. To mitigate non-linearity, an energy calibration must be performed with gamma lines covering the full energy range of interest. Extrapolation outside of the energy range or interpolation between distant energy points is likely to result in significant errors.

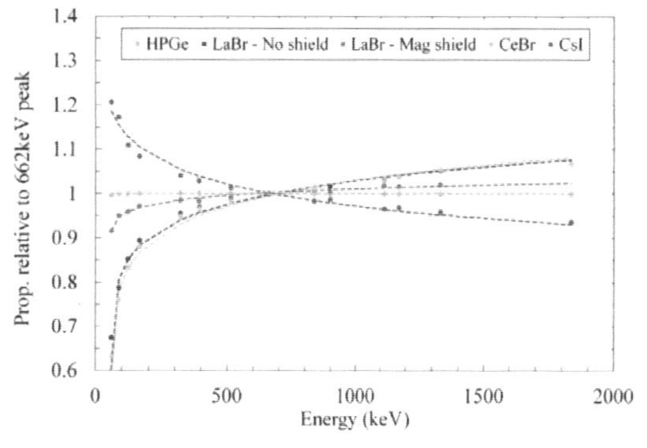

Figure 4. Comparison of detector proportionality as a function of energy. Proportionality defined as channels per keV at a given energy relative to channels per keV at 662 keV. The linearity of the Cd(Zn)Te detector was similar to that of the HPGe detector.

3.3. Internal radioactivity

The background radiation detected by a gamma spectrometer produces interference that can mask the presence of a radionuclide of interest. Typically the background is produced by cosmic-rays and naturally occurring radioactive materials (NORM), such as ^{40}K with an emission at 1461 keV. If the detector material itself contains radioactive material then a significant background will result from the high detection efficiency for internally emitted radiation.

A comparison of the background energy spectra measured by each detector is shown in figure 5. The background of Cd(Zn)Te is the lowest, due to its low absolute efficiency and low internal radioactivity. Several NORM gamma lines are evident in the HPGe spectrum, also suggesting low internal radioactivity. Evidence for alpha-emitting contaminants was found in the $LaBr_3$(Ce) and $CeBr_3$ spectra (between 2,000-2,500 keV for $LaBr_3$(Ce) and between 1,400-2,000 keV for $CeBr_3$). This was due to the presence of ^{227}Ac, which is part of the natural ^{235}U decay series. The contamination of $CeBr_3$ varies from batch-to-batch with certain detectors showing almost no evidence for internal alpha emission [8]. Due to the chemical similarity of La and Ac, ^{227}Ac contamination is harder to remove from $LaBr_3$(Ce). $LaBr_3$(Ce) also contains the naturally abundant radioactive isotope ^{138}La. Evidence was found for this in the $LaBr_3$(Ce) spectrum, in the form of a gamma peak at 1436 keV and gamma/beta sum peak at 789 keV [6]. The benefit of internal radioactivity is the presence of a consistent reference peak for gain stabilisation (which is required for detectors such as $LaBr_3$(Ce)/$CeBr_3$ that are known to drift with temperature).

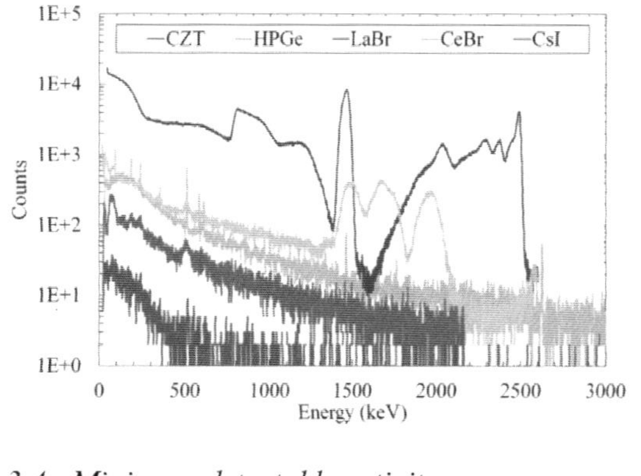

Figure 5. Comparison of background energy spectra measured in Cu/Cd lined Pb coffin over 24 hr.

3.4. Minimum detectable activity

The MDA of a detector is often quoted when a positive identification of a radionuclide is not made. It provides insurance against claiming a radionuclide is not present, simply because it was not detected. An air monitor with large MDA values for key radionuclides is not useful as the operator will be forced to report high concentrations of those radionuclides on a routine basis. Actual releases may well be missed.

MDA values for three key radionuclides were calculated; ^{241}Am, ^{137}Cs and ^{60}Co, and are presented in figure 6. The HPGe and $CeBr_3$ detectors performed best, with MDA values for 24 hr measurements of 0.1 and 0.2 Bq, respectively. It would take just 100 and 200 s, respectively, for these levels of contamination to be collected on a filter in an environment of 0.03 Bq/m^3, even with a modest pump rate of 10 m^3/hr (typical of a handheld air sampler). With this level of performance it would be prudent to better balance the sampling-to-measurement times (if not concurrent). The MDA values of Cd(Zn)Te were found to increase with energy as detection efficiency decreased. The MDA values of $LaBr_3$(Ce) were found to decrease with energy. This can be explained with reference to the internal background (see figure 5), which decreases from 60 to 1332 keV. The MDA values produced by the CsI(Tl) detector were reasonable despite the poor energy resolution.

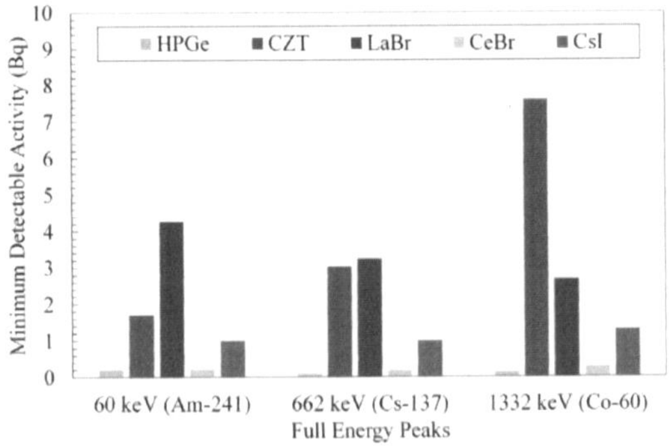

Figure 6. Comparison of MDA values for ^{241}Am, ^{137}Cs and ^{60}Co when measured in Cu/Cd lined Pb coffin over 24hr.

4. Summary and conclusions

A comparison of new and emerging gamma detector technologies for airborne radioactivity measurements has been completed. The key performance metrics of energy resolution, detection efficiency and MDA have been assessed.

The HPGe detector produced superior resolution, efficiency and MDA values compared with the other detectors. However the HPGe detector tested was large, making it impractical to mount within an air sampler unless some form of filter progression was employed. A remote electromechanical cooler would improve practicality. The cost also limits the number of monitors that could be deployed across a network. The Cd(Zn)Te detector demonstrated good potential for radionuclide identification and was the most compact. The poor detection efficiency would however extend measurement times and delay response. The internal radioactivity of $LaBr_3$(Ce) resulted in large MDA values. The significantly lower internal background of $CeBr_3$ compensated for its modestly worse energy resolution. The $CeBr_3$ MDA values were comparable with those of the HPGe detector. The cost of $LaBr_3$(Ce)/$CeBr_3$ (inc. PMT and associated electronics) is typically lower than HPGe so more monitors could be deployed. Despite the poor energy resolution and relatively small volume, the CsI(Tl) detector produced reasonable MDA values. The low cost (over an order of magnitude lower than HPGe) and compact form are both highly desirable when considering in-situ gamma spectrometers for wide-scale deployment.

The detectors investigated covered the full range of the price-performance matrix and the results of the comparison will assist instrument manufacturers and networks operators in determining which technology satisfies their individual requirements.

Acknowledgments

The authors would like to acknowledge funding by the Department of Business Innovation and Skills through the National Measurement System, Euramet through the EMRP programme, the STFC though grant #ST/L005743/1 and the NDA through a PhD bursary. The authors would like to acknowledge Kromek Group PLC for the loan of detectors and support of the project. PHR acknowledges support from the STFC (UK) via grant number ST/L005743/1.

References

[1] Treaty establishing the European Atomic Energy Community http://ec.europa.eu/energy/en/topics/nuclear-energy/ last accessed 05/08/16.

[2] Cindro D *et al.* 2016. http://www.nss.si/nene2016/contents.html#5300 last accessed 05/06/16

[3] Navarro Ortega N *et al.* 2015 https://www.ctbto.org/fileadmin/user_upload/SnT2015/SnT2015_Posters/T3.2-P4.pdf last accessed 05/08/16.

[4] MetroERM Consortium http://earlywarning-emrp.eu/ last accessed 05/06/16.

[5] Currie L A 1968 *Anal. Chem.* **40** (3) pp 586–93 DOI: 10.1021/ac60259a007.

[6] Lorusso G, Shearman R, Regan P H, Judge SM, Bell S, Collins S M, Larijani C, Ivanov P, Jerome SM, Keightley JD, Lalkovski S, Pearce AK, Podolyak Zs., 2016 Applied Radiation and Isotopes **109** pp507-511 DOI: http://dx.doi.org/10.1016/j.apradiso.2015.12.050

[7] Commission recommendation 2000/473/Euratom http://eur-lex.europa.eu/legal-content/EN/TXT/HTML/?uri=CELEX:32000H0473&from=NL last accessed 05/06/16.

[8] Quarati F G A 2013 *NIMA* **729** pp 596-604 DOI: 10.1016/j.nima.2013.08.005.

Journal of Physics: Conference Series **763** (2016) 012011 doi:10.1088/1742-6596/763/1/012011

Scintillators and Cherenkov detectors for the registration of 10.8 MeV gamma rays

J E McMillan[1], **A Cole**[1], **A Kirby**[1] **and M L Ahnen**[2]

[1] Department of Physics and Astronomy, The University of Sheffield, Sheffield, South Yorkshire, S3 7RH, UK

[2] ETH Zürich, Institute for Particle Physics, Otto-Stern-Weg 5, 8093 Zürich, Switzerland

E-mail: j.e.mcmillan@sheffield.ac.uk

Abstract. The identification of nitrogen by neutron activation has been utilized in both explosive detection and in-vivo metabolic analysis. The 10.8 MeV gamma ray line emitted by thermal neutron capture provides a unique signature, however, due to its high energy its registration is non-trivial. Conventional approaches have used large dense inorganic scintillators which inevitably entail considerable expense. We examine the capabilities of arrays of smaller scintillation detectors and the use of glass Cherenkov detectors as an alternative.

1. Introduction

High energy gamma rays are produced as a result of neutron activation of materials. There is particular interest in detecting the 10.8 MeV gamma-ray line produced as a result of thermal capture on nitrogen. The cross-section for the capture reaction on nitrogen is $\sigma_{\gamma,\mathrm{all}} = 79.0$mb while the component generating 10.8 MeV gamma-rays is $\sigma_{\gamma,10.8} = 11.3$mb. This reaction has the potential for use as an indicator of the presence of explosives, since all commercial and most improvised explosives are rich in nitrogen [1]. There has also been some work on in-vivo nitrogen/carbon ratios as a diagnostic of human metabolic disorders [2].

Few elements, and certainly no other common element, has significant gamma emission at 10.8 MeV or higher [2]. While the detection of these high energy gamma-rays offers a unique signature, actual detection is more problematic, due to the size and density of detectors required for total absorption. The task is made more difficult by the fact that the gamma detectors are inevitably operated in a mixed field also containing neutrons of all energies from production down to thermal.

High-purity germanium (HPGe) detectors are ill-suited to the task, having low efficiency and requiring cryogenic cooling. Germanium suffers from activation problems [3], while the expense of HPGe and other semiconductor detectors rules them out for deployment in field deployed explosive detector systems.

Conventional wisdom is to use large dense inorganic scintillators, and the obvious choice is sodium iodide (NaI(Tl)). Data from Saint-Gobain [4] indicate that at 10.8 MeV, 50mm NaI(Tl) crystals have an absorption efficiency of 50% and that 120mm crystals have an efficiency of around 80%. The studies by Stamatelatos et al.[3], McFee et al.[5] and Gierlik et al.[6] show that there is little to be gained by using recent dense inorganic scintillators such as bismuth germanate (BGO) or lanthanum bromide ($LaBr_3$). These materials have activation problems

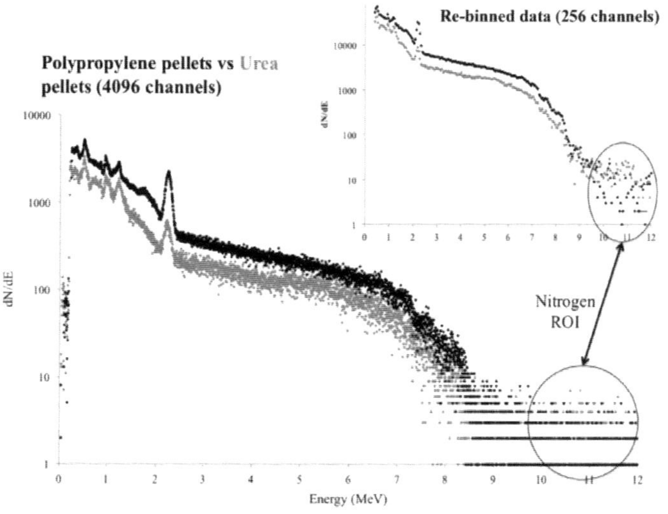

Figure 1. Gamma spectrum produced on 10kg urea sample in 300 minutes.

and the improvements in detection efficiency are outweighed by the huge increase in cost over NaI(Tl). Overall detection efficiency can be improved more cost-effectively by deploying a larger area of NaI(Tl).

Stamatelatos et al.[3] emphasize that the measurement is a signal-to-noise problem and that reducing background is more important than increasing signal. They stress that, in the environment of a neutron source, it is necessary to shield the detectors adequately from both gamma-rays coming from directions other than the target and from neutrons. They suggest the use of bismuth for the gamma shielding as it activates to a much lesser extent than lead and the resulting activation products are not gamma emitters. Further they suggest a layered shielding approach with both neutron and gamma absorbing materials. Closest to the crystal, they used a layer of resin loaded with ^{6}LiF, to absorb thermal neutrons without the emission of gamma-rays. ^{6}LiF is now too expensive to be considered for shielding, but we have found that a 10mm layer of a 40/60 mixture of wax and unenriched lithium carbonate intended for ceramic glazes is similarly effective at suppressing thermal neutrons.

Stamatelatos et al.[3] also note that significant background is caused by pulse pile-up in which two lower energy gamma-rays interact in the crystal within the decay time of 250ns. They point out that this can be reduced by using a number of crystals 50mm diameter by 100mm long, each with its own readout, rather than a large single crystal.

2. Measurements with DT generator

Using an NSD-Fusion 14 MeV DT neutron generator and pelleted urea as an explosives surrogate, we have demonstrated this technique [7]. The emitted gamma-rays were recorded with a single $4'' \times 4''$ NaI crystal fed to a pulse height analyzer. The crystal was lead shielded in all directions but the forward. The pulses were time gated in order to be optimally sensitive to activation caused by thermal neutrons. Typically three hour exposures were employed on 10kg urea samples which provided a sufficiently detailed spectrum to understand the processes.

A typical observed spectrum is shown in figure 1. The background caused by pulse pile-up is particularly severe in the region 4–7 MeV. By subtracting the background polypropylene spectrum from the urea spectrum, the 10.8 MeV emission can clearly be seen as in figure 2. Despite the pulse pile-up and broadening which it causes, a statistically significant peak could

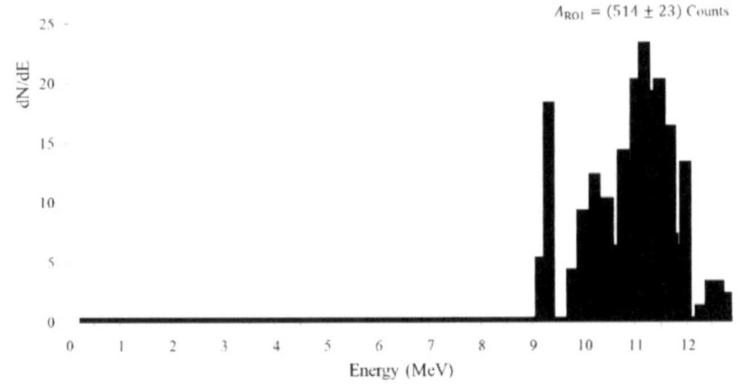

Figure 2. 10.8MeV peak obtained from 10kg urea sample in 300 minutes

be obtained in around ten minutes. Further details of these measurements can be found in Cole [7].

3. Cherenkov detectors

To achieve similar results in shorter time, it is clear that a larger area of detector must be used. As arrays of NaI(Tl) scintillators are likely to remain expensive, it was decided to investigate whether much cheaper Cherenkov detectors could be used.

Cherenkov detectors with dense radiators have found extensive use as charged particle detectors but are also useful for gamma-ray detection where fast timing is required and where good energy resolution is not important. Sowerby [8] examined acrylic Cherenkov detectors for use over the range 1.1–2.6 MeV and Sober et al. [9] consider lead-glass detectors of 153mm depth for the region 10-20 MeV. They show there is a threshold below which no signal is produced, while at 10.8 MeV the efficiency can be expected to be around 60%. Grannis et al. [10] give indications regarding low-cost production of these detectors. Ayaz-Maierhafer et al. [11] have recently demonstrated the utility of glass Cherenkov detectors in the detection of fission gammas. Ayaz-Maierhafer et al. [12] have also performed experimental measurements on neutron activation of Cherenkov glasses which suggest that their long term operation in mixed neutron-gamma fields is not likely to be problematic. Hayward et al. [13] provide a detailed simulation of the response of glass Cherenkov detectors to 6 MeV gamma rays.

It is evident from these studies, that 10.8 MeV is at the lower end of the useful energy range of Cherenkov detectors. However, they offer large size, high density and consequently high efficiency at low cost. Lead-glass is inherently cheaper than NaI(Tl), as an indication, Schott F2 glass was quoted as £56/kg [14]. Being non-hygroscopic, no special precautions are needed in construction and the resulting detectors are robust. Their time response is limited by the risetime of the photomultiplier used and is typically 10ns, so problems with pulse pile-up are much reduced.

The obvious disadvantage of Cherenkov detectors derives from the low light output and the subsequent poor energy resolution at 10.8MeV. In the absence of distinguishable peaks in the output spectra, the calibration of Cherenkov detectors is also problematic. This is compounded by the unavailability of sources with suitable high-energy emission. Calibration has been considered in detail by Rose and Erickson [15].

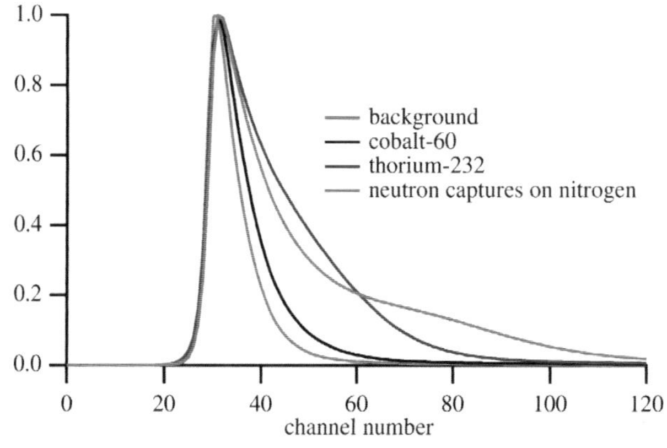

Figure 3. Normalized pulse height spectra obtained from lead-glass detector

3.1. Experimental detector

To test these ideas, an experimental lead-glass Cherenkov detector was constructed. A block of Schott F2 lead-glass 80mm×80mm×80mm was obtained. The surfaces of this were shot-blasted, with the exception of a circular output window on one face. The block was then painted with Edmund 53-697 $BaSO_4$ high reflectance paint, which provides good diffuse reflectance, even in the ultraviolet. The block was then coupled to an ET9265K 70mm diameter photomultiplier using a transparent silicone pad. The whole assembly was made light-tight with aluminium foil and black tape.

The resulting detector worked largely as expected. The 9265K has a rise-time of 7.5ns, so the output pulses are around 15ns FWHM. Figure 3 shows normalized pulse height spectra obtained in response to different sources. There is clearly an offset of some 20 channels on the x-axis, which should be ignored.

The spectrum taken with ^{60}Co, having peaks at 1.1 and 1.3 MeV shows little difference from background. The ^{232}Th spectrum was taken with an old sample of ThO, so had emissions from all the progeny, of which ^{208}Tl is the most important as it decays with a major branch emitting a gamma of 2.62 MeV. The neutron captures on nitrogen spectrum was obtained by placing a few kilograms of urea in front of the detector and exposing it to thermal neutrons obtained from an AmBe source in a mass of polypropylene moderator. It can be seen that by channel 60, background and low-energy gammas are negligible. Any events registered with a height greater than channel 75 are interesting and indicative of high-energy gamma emission.

Further to the spectra shown, there was a population of much larger pulses, with a peak at around channel 1200 and a high energy tail. The peak was attributed to single cosmic muons traversing the block, while the higher energy events are probably showers and residual gas events in the photomultiplier. As the rate of these events was many orders of magnitude less than that expected in a screening scenario, they can be ignored.

These results are broadly similar to those obtained by Sober et al. [9] and Rose and Erickson [15]. While detailed measurements have not yet been made, it appears that our experimental detector has an electron detection threshold at ~138keV and at 10.8 MeV, a detection efficiency of ~50% and an energy resolution of ~50%.

4. Detector arrays

The object of the proposed measurements is not to produce a detailed emission spectrum of the sample being irradiated, but simply to establish whether the emission is consistent with the presence of emission at 10.8 MeV. Consequently the analysis of peaks in regions-of-interest can be replaced with the simple registration of Cherenkov signals exceeding a threshold.

A more convincing result may be obtained by combining the signals from a small area of NaI(Tl) detectors with a larger area of Cherenkov detectors. An array containing a mixture of the two types of detectors is envisaged. In the presence of 10.8 MeV emission the NaI(Tl) would record a small number in the relevant peak region-of-interest, while the Cherenkov detectors would record a large number of high-energy events. The most obvious conclusion would be that the Cherenkov events have the same energy distribution as those recorded by the NaI(Tl) detector. The presence of high-energy Cherenkov events with no signal in the NaI(Tl) region of interest would be indicative of neutron activation of some other element than nitrogen.

5. Combining datasets

The combination of data from detectors with different resolutions and efficiencies and the determination of the presence of 10.8 MeV gamma emission can be examined using objective Bayesian analysis [16, 17]. An algorithm using Bayes Factors calculated from count rates over specified periods for the two types of detector is proposed. In the case of the NaI(Tl) detector there is a region of interest (ROI) which includes the 10.8 MeV peak, while in the case of the Cherenkov detector, the high-energy region of interest is everything above a discrimination threshold. The Bayes factors B_{01} are calculated as per equation 3.1 of [17]. Similarly the threshold is set as a Bayesian z-value as per equation 3.3 of [17].

Record background for a certain time t_{off}:
  Record background with NaI detector,
    number of events in the 10.8MeV RoI is $N_{off,\mathrm{NaI}}$
  Record background with Cherenkov detector,
    number of events in the high-energy RoI is $N_{off,\mathrm{Che}}$

Record sample for a certain time t_{on}:
  Record sample with NaI detector,
    number of events in the 10.8MeV RoI is $N_{on,\mathrm{NaI}}$
  Record sample with Cherenkov detector,
    number of events in the high-energy RoI is $N_{on,\mathrm{Che}}$

Calculate individual Bayes Factors that events are due to background only:
  Calculate Bayes Factor for NaI detector with:
    $B_{01,\mathrm{NaI}}(N_{on,\mathrm{NaI}}, N_{off,\mathrm{NaI}}, t_{on}/t_{off})$
  Calculate Bayes Factor for Cherenkov detector with:
    $B_{01,\mathrm{Che}}(N_{on,\mathrm{Che}}, N_{off,\mathrm{Che}}, t_{on}/t_{off})$

As Bayes Factors are commutative, the experiments can be performed in any order.
The new odds result from previous odds by multiplication:

Resulting odds of the background (H_0) over signal (H_1) hypothesis:
  $P(H_0)/P(H_1) = B_{01,\mathrm{NaI}} * B_{01,\mathrm{Che}}$

Set a threshold. If $P(H_0)/P(H_1) < 5.7 \times 10^{-7}$ then there is 5σ detection

---

6. Current status

The shielding of both the NaI(Tl) and Cherenkov detector needs improvement and more detailed calibration runs of the Cherenkov detector are planned. The Bayesian algorithm is being tested with simulated data. If the initial experiments look fruitful, a larger detector array will be built.

Acknowledgements

This work was supported by EPSRC EP/G005141/1 "Neutron Screening for Airborne Cargo". We would like to thank Prof. Dick Lacey of the Home Office Centre for Applied Science and Technology and Dr. Marieke Navin for the fabrication of the lead-glass block.

References

[1] Litz M *et al.* 2012 Neutron-activated gamma-emission: Technology review Tech. Rep. ARL-TR-5871 Army Research Laboratory, Adelphi, MD, USA
[2] Molnar G 2004 *Handbook of prompt gamma activation analysis* (Kluwer Academic Publishers, Dordrecht, NL)
[3] Stamatelatos I, Chettle D and Scott M 1993 *Phys. Med. Biol* **38** 411
[4] 2014 Efficiency calculations for selected scintillators, Saint-Gobain Ceramics & Plastics, Inc.
[5] McFee J E *et al.* 2009 *IEEE Trans. Nucl. Sci.* **56** 1584–1592
[6] Gierlik M *et al.* 2012 *IEEE Trans. Nucl. Sci.* **59** 230–235
[7] Cole A 2015 *Installation and Characterization of a pulsed-neutron generator at the University of Sheffield Neutron Facility for use in a Feasibility Study into Pulsed Fast-Thermal Neutron Activation Analysis for Cargo Screening* Ph.D. thesis The University of Sheffield
[8] Sowerby B 1971 *Nucl. Instr. Meth.* **97** 145–149
[9] Sober D *et al.* 1973 *Nucl. Instr. Meth.* **108** 573–579
[10] Grannis P, Jaffe D and Marx M 1981 *Nucl. Instr. Meth. Phys. Res.* **188** 239–242
[11] Ayaz-Maierhafer B, Laubach M A and Hayward J P 2016 *J. Radioanal. Nucl. Chem.* **308** 919–926
[12] Ayaz-Maierhafer B *et al.* 2013 *IEEE Trans. Nucl. Sci.* **60** 701–707
[13] Hayward J P *et al.* 2013 *J. Radioanal. Nucl. Chem.* **295** 1321–1329
[14] 2016 David Bentley, private communication, H.V.Skan Ltd.
[15] Rose P and Erickson A 2015 *Nucl. Instr. Meth. Phys. Res.* **A799** 99–104
[16] Knoetig M 2014 *Astrophys. J.* **790** 106 (8pp.)
[17] Ahnen M 2015 *34th International Cosmic Ray Conference, POS(ICRC2015)701*

Journal of Physics: Conference Series **763** (2016) 012012 doi:10.1088/1742-6596/763/1/012012

A scintillating fibre beam profile monitor for the experimental areas of the SPS at CERN

I Ortega[1], **A Bay**[2], **G Haefeli**[2], **J Spanggaard**[1], **G Tranquille**[1]

[1] European Organization for Nuclear Research (CERN), Geneva, Switzerland
[2] Ecole Polytechnique Fédérale de Lausanne (EPFL), Lausanne, Switzerland

E-mail: inaki.ortega@cern.ch

Abstract. The CERN Super Proton Synchrotron (SPS) delivers a wide spectrum of particle beams (hadrons, leptons and heavy ions) that can vary greatly in momentum and intensity. The profile and position of these beams are measured using particle detectors. However, the current systems show several problems that limit the quality of such monitoring. We have researched a new monitor made of scintillating fibres read-out with Silicon Photomultipliers (SiPM), which has the potential to perform better in terms of material budget, range of intensities measured and available detector size. In addition, it also has particle counting capabilities, extending its use to spectrometry or Time-Of-Flight measurements. Its radiation hardness is good to guarantee years of functioning. We have successfully tested a first prototype of this detector with different particle beams at CERN, giving accurate profile measurements over a wide range of energies and intensities. It only showed problems during operation with lead ion beams, believed to come from crosstalk between the fibres. Investigations are ongoing on alternative photodetectors, the electronics readout and solutions to the fibre crosstalk.

1. Introduction

In the experimental areas of the SPS, protons are extracted during 4.8 seconds and collided with primary, secondary and sometimes tertiary targets, in order to produce beams of particles that can be selected and sent to the experimental users. These beams can be composed of hadrons (protons, kaons, pions, antiprotons...), leptons (electrons, positrons, muons...) and lead ions. Their momenta can vary greatly, from 1 to 400 GeV/Z/c, and their intensities from 10^3 to 10^8 particles per second. The profile and position of these beams are typically measured using particle detectors based on gaseous active materials (multi-wire proportional chambers (MWPC), delay wire chambers (DWC) and gas electron multipliers (GEM)) or scintillators (finger scan scintillators (FISC)). Replacement detectors for wire chambers are actively being sought as they are ageing and the expertise to produce them is gradually being lost.

In addition, two new beam lines dedicated to neutrino R&D will be commissioned in 2017, in collaboration with Fermilab and other institutes. These new lines will provide tertiary beams of hadrons and leptons with low energies (12 GeV down to 500 MeV), intensities of maximum 10^3 particles per second and a beam spot of several cm in diameter. The monitors for these lines will form a spectrometer for particle momentum measurement and therefore need to count single particles, while the large spot size requires covering an area of $200 \times 200\,\mathrm{mm}^2$. This implies that the current monitors cannot be used in those beam lines: the MWPC and the FISC cannot provide single particle counting; the DWC, which can count individual particles is too small ($100 \times 100\,\mathrm{mm}^2$); the GEMs are too thick in material budget for the low energy beams.

2. Scintillating fibres

Scintillating plastic fibres (SciFi) have emerged as one of the best active materials for the monitors of the experimental areas. They have a core made of polystyrene cladded with one or two layers of lower refractive index material. This gradient of refractive index allows a fraction of the light created inside the fibre to be trapped by total internal reflection. The polystyrene fibre core usually employs a two level doping system: a primary scintillator emitting in the UV and a wavelength shifter to capture the short reach UV photons and re-emit them in the visible wavelength region. This shift in wavelength also enhances the match in terms of quantum efficiency for common photodetectors. The processes of energy absorption, scintillation and wavelength shifting are mediated by very fast quantum processes that yield a photon time distribution with rise time and decay time of 1-3 ns [1]. Light production typically reaches up to 8000 photons per MeV of energy deposited, although the trapping efficiency of square fibres varies between 4.2% and 7.3% [2][3]. Depending on the amount of dopants, the light emitting properties of the fibres can be changed and their radiation hardness can be improved.

The two main manufacturers of scintillating fibres, Kuraray and Saint-Gobain, provide fibres with similar properties. The emission spectra from both peak between 430-450 nm, but the light yield and attenuation length differ, most likely due to differences in manufacturing, doping concentrations and scintillators used.

2.1. Radiation hardness

A very important characteristic of the detector is its radiation hardness. The detector should be able to operate continuously and reliably for years with beams of intensities of 10^8 particles/second or 10^6 Pb ions/s. The radiation damage is mainly manifested as a shorter attenuation length of the fibre resulting in less light collected by the photodetectors [4]. Short fibres are less susceptible to the effects of radiation damage with data from literature showing that short fibres can withstand doses of up to 10 kGy before showing significant damage [5].

Simulations of a SciFi detector carried out with Geant4 [6] show that for a single beam extraction of 10^8 particles, an absorbed dose of 100 mGy can be expected per fibre. Such short fibres should therefore withstand up to 10^5 of such beam extractions, guaranteeing several years of operation before having to be replaced.

2.2. Material budget

It is important for a beam monitor to perturb the measured beam as little as possible. A charged particle traversing a medium is deflected due to Coulomb scattering from nuclei, characterized by the radiation length X_0 and the nuclear interaction length λ. According to the multiple scattering theory of Molière [7], the angular distribution of the scattered particles has a root mean square (*r.m.s.*) given by:

$$\theta_0 = \frac{13.6\,\mathrm{MeV}}{\beta c p} Z \sqrt{x/X_0}\left[1 + 0.038\ln(x/X_0)\right]. \quad (1)$$

Where x is the thickness of the material, Z the charge of the particle, p its momentum and βc its speed. We have studied the x/X_0 of different monitors to establish a comparison between them. As shown in Table 1, a SciFi monitor composed of two planes (horizontal and vertical) of 0.5mm square fibres has a material budget slightly below the current monitors.

Table 1. Comparison of x/X_0 for different monitors

Detector	x/X_0 (%)
MWPC	0.34
DWC	0.25
SciFi 1 mm	0.47
SciFi 0.5 mm	0.24

3. First prototype

A first detector was built and tested in the H8 beam line of the SPS North Experimental Area at CERN. It was built with only one plane for simplicity. This plane is composed of 64 Saint-Gobain BCF-12 scintillating fibres of 1 mm thickness and square shape. The length of the fibres is 35 cm and they are packed together along one row, leaving no space between them. As the electronics readout chip used in this detector only has 32 channels, it was decided to read-out only every other fibre. This gives a spatial resolution of 2 mm within an active area of $64 \times 64\,\mathrm{mm}^2$. A mirror is glued on the end of the fibres to increase the light collection.

The scintillating fibre detector can be operated in vacuum, avoiding additional vacuum windows further decreasing the total material budget. A fibre feedthrough, sealed with an Epoxy glue (Araldite Standard) suitable for the primary vacuum of the North Area beam lines, was used to allow installation of the photodetectors outside the vacuum tank.

3.1. Photodetectors

The photo detector chosen to read the light from the scintillating fibres were Silicon Photomultipliers (SiPM), specifically the Multi-Pixel Photon Counter (MPPC) model S13360-1350 from Hamamatsu [8] (although other brands (SensL and KETEK) were tested in the laboratory). These silicon devices show a very good photo-detection efficiency of 40% in the relevant light-emitting wavelength range. They have high gain ($\approx 10^6$) and provide fast pulses with sub-nanosecond rise time and 50-100 ns fall time.

Other advantages are their compact size, their insensitivity to magnetic fields and their low operating voltage ($< 100\,\mathrm{V}$). Their most important drawback is the high dark count rate (typically $\approx 100\,\mathrm{kHz/mm}^2$), which can have a thermal origin (due to its semiconductor nature) or can be caused by after-pulses generated some nanoseconds after a true event. These detectors also show crosstalk between pixels, resulting in artificially bigger pulses. The crosstalk in the first versions was as high as 10%, but more recent devices, like the S13360-1350 MPPC, show a much lower crosstalk (1%) and less after-pulsing.

An alternative to read multiple scintillating fibres efficiently are Multi-Anode Photomultipliers (MA-PMT). We tested the Hamamatsu H7546 [9], which has 64 channels over an active area of $18.1\,\mathrm{mm} \times 18.1\,\mathrm{mm}$. It shows a good quantum efficiency ($\approx 35\%$), large gain ($\approx 10^6$), fast rise and fall times of 1-2 ns and 2-3 ns respectively and a lower dark count rate than the SiPM ($\approx 100\,\mathrm{Hz}$). However, MAPMT have problems of gain uniformity and crosstalk between channels. For the first prototype we favoured the SiPM as they were considered to be a new technology with a big margin for improvement and potentially lower future production costs.

3.2. Electronics readout

The analogue pulses from the SiPMs were processed by the CITIROC ASIC, developed by OMEGA Microelectronics [10]. This chip allows amplification, discrimination and integration of 32 pulses simultaneously. Another interesting feature is a fine-tuning of the SiPM voltages, allowing gain equalization, as each individual SiPM requires a slightly different operating voltage in order to achieve a homogenous detector response.

The CITIROC ASIC also has a trigger line composed of a fast shaper ($\approx$ 15 ns) and a discriminator, which produces a logical signal whenever the incoming pulse exceeds a pre-set threshold. Thanks to the low cross talk of the MPPC used ($< 1\%$), the spurious noise pulses rarely exceeded the four photon level, whilst the light detected in the 1 mm fibres for Minimum Ionizing Particles (MIPs) follows a Landau distribution with a most probable value of thirty photons. It was therefore considered safe to set the discrimination threshold to four photons.

3.3. Data acquisition

The logical pulses were sent to VME Scalers, where every channel corresponds to one of the fibres read in the detector. A LabVIEW module was used to read the Scalers and present and store the data on a PC. This data acquisition system was synchronized with the timing signals coming from the SPS machine in such a way that the Scalers were reset immediately before the beam extraction and their data read immediately after the extraction finished.

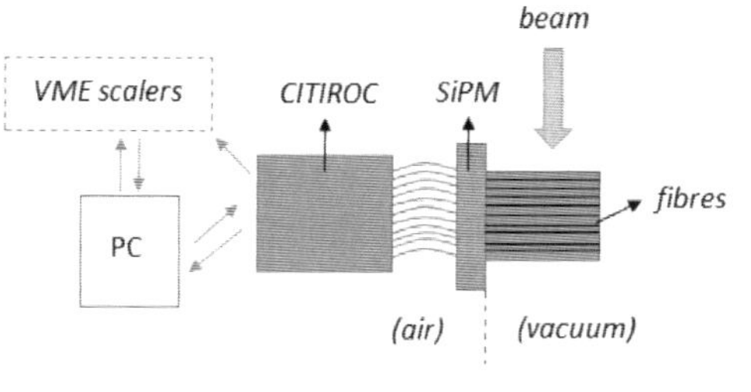

Figure 1. Schematic of the detector showing its main parts

4. Beam tests in H8

The prototype was installed close to two other profile monitors: a DWC placed upstream and a FISC downstream. This allowed direct comparison between them. A scintillator counter placed upstream provided accurate intensity measurements.

From the 20th of October to the 11th of November of 2015, we monitored hadrons and leptons with momenta between 20 GeV/c and 180 GeV/c and intensities from 10^3 to 10^6 particles/second. The profiles have been analysed with Root [11] to fit a Gaussian curve and find the *r.m.s.*. In the following figures (2, 3 and 4) we show some of the profiles seen by the SciFi, the DWC and the FISC of beams of protons mixed with pions of momenta 180 GeV/c and different intensities. An analysis of these profiles is shown in Table 2.

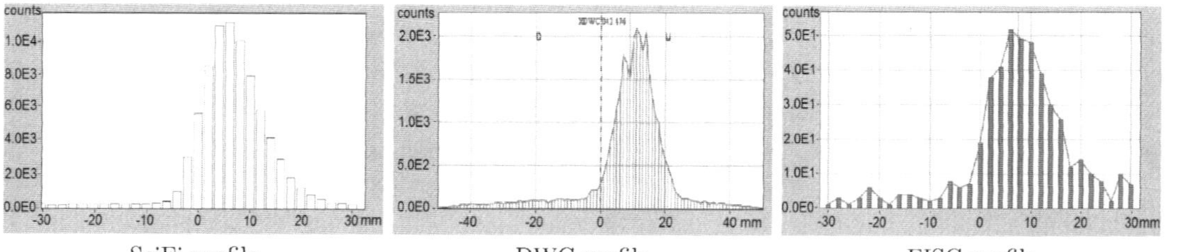

SciFi profile DWC profile FISC profile

Figure 2. Profiles of a 180 GeV/c proton/pion beam of $I = 3.4 \times 10^4$ particles/s

Journal of Physics: Conference Series **763** (2016) 012012 doi:10.1088/1742-6596/763/1/012012

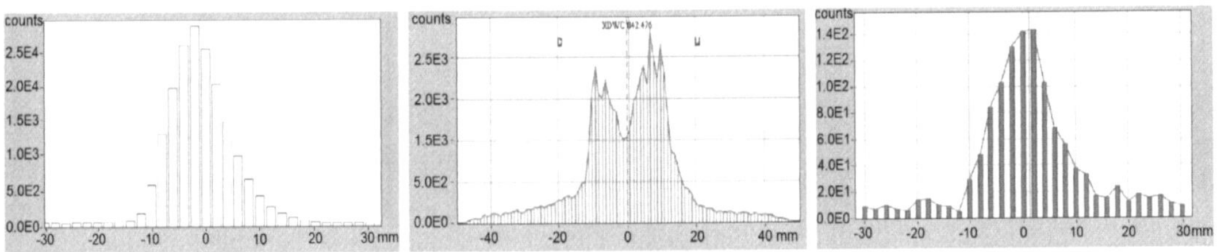

SciFi profile DWC profile FISC profile

Figure 3. Profiles of a 180 GeV/c proton/pion beam of $I = 8.2 \times 10^4$ particles/s

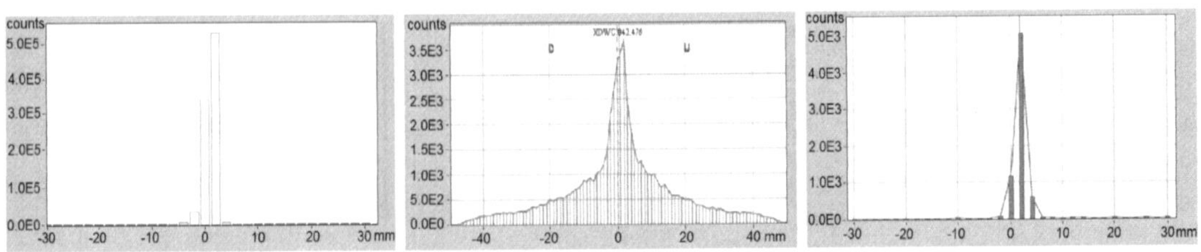

SciFi profile DWC profile FISC profile

Figure 4. Profiles of a 180 GeV/c proton/pion beam of $I = 6.5 \times 10^5$ particles/s

Table 2. Comparison of the *r.m.s.* of the previous figures shown (2, 3 and 4)

Intensity(particles/s)	σ(mm) SciFi	σ(mm) DWC	σ(mm) FISC
3.4×10^4	5.6	5.8	6.6
8.2×10^4	5.4	11.2	6.2
6.5×10^5	0.9	4.0	1.1

The SciFi monitor worked satisfactorily in all situations, whilst the DWC had troubles with the high intensities, showing distorted profiles or artificial tails; the FISC on the other hand was unable to work at intensities lower than 10^4 particles/s. In addition to that, the intensity measured by the fibres agreed with the intensity from the scintillation counter.

4.1. Lead ion run

From the 18th of November to the 30th of November of 2015, the SPS cycle changed to lead ions, $^{208}Pb^{+82}$, providing beams of these heavy particles directly to the North Area. Lead ions deposit four orders of magnitude more energy than MIPs, which means that the light produced and collected also grows by four orders of magnitude. It was therefore necessary to lower the operating voltage of the SiPM to decrease the photon detection efficiency and avoid saturation.

As seen in Table 3, for lead ions the profiles from the SciFi were seen to be wider than those from the DWC, in particular for high intensity beams (comparison with the FISC was unfortunately not possible as they did not work properly with the Pb ions due to technical problems). We believe that the origin of these wider profiles in the SciFi are due to crosstalk between the fibres. This crosstalk is caused by primary UV photons created during the scintillation that can escape the fibre and travel to neighbouring fibres, where they excite the wavelength shifting dopants. Because of the larger energy deposition from Pb ions, a much larger number of these crosstalk photons are created, explaining the wider profiles. This could be avoided in the future by treating the fibre cladding with a UV absorber or reflector.

Table 3. Comparison of the *r.m.s.* of different Pb ion beam profiles

Intensity(particles/s)	σ(mm) SciFi	σ(mm) DWC
3.7×10^2	5.5	4.2
2.4×10^4	7.7	7.0
1.0×10^6	9.6	5.1

5. Conclusions and prospects

A scintillating fibre monitor has been successfully tested in the H8 beam line at CERN where it has shown that it can replace the existing beam monitors over a wide range of intensities, presenting less material for the beam and giving more accurate profiles.

A second prototype using MAPMT instead of SiPM for light detection has now also been built to allow both technologies to be compared before deciding on the final design. In addition, two new versions of the detector have been built to investigate solutions to the crosstalk: one detector has the fibres untreated, whilst the fibres of the other are coated with an ultra-thin aluminium layer following the example of ATLAS ALFA [12]. This aluminium coating will act as a reflector for the ultra-violet photons coming from outside the fibre.

New front-end electronic boards will also be tested, replacing the VME Scalers to allow tagging events with both spatial and time information. This will enable the possibility of reconstructing both the transverse and longitudinal profiles of the beam.

References

[1] White T O 1988 *Nucl. Instr. Meth. Phys. Res. Sect. A* **273** 820–825
[2] Kuraray Co., Ltd. *Plastic scintillating fibres* product catalogue http://Kuraraypsf.jp/
[3] Saint-Gobain Crystals *Scintillating Optical Fibers* product catalogue www.crystals.saint-gobain.com
[4] Acosta D *et al.* 1991 *Nucl. Instr. Meth. Phys. Res. Sect. B* **62** 116–132
[5] Jakobsen S 2013 *Commissioning of the absolute luminosity for ATLAS detector at the LHC* Ph.D. thesis The Niels Bohr Institute, Faculty of Science, University of Copenhagen
[6] Agostinelli S *et al.* (GEANT4 Collaboration) 2003 *Nucl. Instr. Meth. Phys. Res. Sect. A* **506** 250–303
[7] Olive K A *et al.* (Particle Data Group) 2015 *The review of Particle Physics* vol 38 (Chin. Phys. C)
[8] Hamamatsu Photonics K.K. *MPPC (multi-pixel photon counter) S13360-1350* datasheet www.hamamatsu.com
[9] Hamamatsu Photonics K.K. *Multianode photomultiplier tube assembly H7546* datasheet www.hamamatsu.com
[10] Fleury J, Callier S, de La Taille C, Seguin N, Thienpont D, Dulucq F, Ahmad S and Martin G 2014 *J. Instrum.* **9**
[11] Brun R and Rademakers F 1997 *Nucl. Instr. Meth. Phys. Res. Sect. A* **389** 81–86
[12] Anghinolfi F *et al.* (ATLAS ALFA collaboration) 2008 Alfa forward detectors for measurement of elastic scattering and luminosity Tech. rep. ATLAS-TDR-18, CERN-LHCC-2008-004, CERN, Geneva.

Acknowledgments

The authors wish to acknowledge Thomas Schneider (CERN EP-DT) for his help with the first laboratory experiments and coating of the fibres. We also thank Lau Gatignon (CERN EN-EA), Cristian Joram (CERN EP-DT), Raphael Dumps (CERN EP-DT) and Sune Jakobsen (CERN BE-BI) for their good advice and support. Special thanks are due to Bernard Florian (EPFL LPHE) and the whole EPFL LPHE workshop for their support building the first prototype.

AUTHOR INDEX

Institute of Physics
Dirac House, Temple Back
Bristol BS1 6BE UK

ISSN: 1742-6588
ISBN 978-1-5108-3181-0

CURRAN ASSOCIATES INC.
proceedings.com

IEEE Radio and Antenna Days of the Indian Ocean (IEEE RADIO 2016)

IOP Conference Series: Materials Science and Engineering
Volume 198

Reunion Island, France
10 – 13 October 2016

IOP
Institute of Physics